Secrets of Outstanding
Site Management

Pro... ...m for

D... ...ces

Ref... ...**jects**

Viktar Zaitsau
CEng, MCIBSE

First Published in Great Britain 2015

by Ecodo Ltd www.EnergyPerformanceHub.com ©

ISBN: 978-0-9576410-2-0

Contents

Part 1: Set the project up for success

1   Why this book? 7

2   How retrofit and refurbishment is different from the new-build construction 8

3   What qualifies as a building services retrofit, what drives the scope of works and the importance of job specifications 10

4   19 expensive mistakes to avoid when pricing the job 13

5   Contracts: protecting yourself against unreasonable client and making your subcontractors perform successfully 23

6   Dilapidation surveys 28

7   Crucial step. How to select star subcontractors and sub-subcontractors 30

8   Cascade your quality and H&S systems throughout the supply chain an reinforce it in the contracts 32

9   Project program. To succeed, account for failures 33

Part 2: Pre-construction

10   Have fun being the principal contractor 39

11   H&S management – don't find yourself in prison 42

12   Finish the design, then build 47

13   Design development 1: schematics and layout drawings 52

14   Design development 2: technical submittals, RFIs, description of operation and installation drawings 53

15   Selecting and buying equipment with a long delivery time 55

16   Shine and glow. Tips for project managers 57

Part 3: Construction

17   Installation program 63

18   Arranging systems shutdowns (micro schedules) 66

19   Electrical and mechanical isolations 69

20   Area acceptance document 71

21   Supervision, labour, then management 73

22   Site set up, mess facility and storage 75

23 Prefabrication and off-site manufacturing 77

24 Site induction – set the rules straight 79

25 Dealing with accidents. Witnessing statements, investigations, actions and toolbox talks 81

26 Balance between site and off site work 83

27 Biggest showstoppers during the installation – the baker's dozen 85

28 Managing the customer 92

29 What are meetings for? 94

Part 4: Special bonus feature. Guaranteed system for managing your project successfully

30 Guaranteed system for managing your projects 99

Part 5: Closing the project

31 Completing your first area: how to snag 109

32 Certification and commissioning 111

33 Six ways subcontractors cut corners without you noticing it 114

34 Site instructions, change of scope and variations 118

35 Recovery of works and acceleration 120

36 How practical completion is unpractical 122

37 O&Ms - paperwork or a building guide? 125

38 Training 127

39 Up selling additional works, becoming a trusted partner and preferred supplier 129

Afterword 131

Appendix A 133

PART 1

Set the project up for success

Why this book?

I am going to give you an *enjoyable* and guaranteed *system* of delivering *successful* building services projects. This system will give you a *feeling* of *masterfully gliding* through the construction project, delivering it *smoothly*, almost *effortlessly* and *fully satisfying* yourself, your boss and customers. Your client will be *delighted* with you.

From the outside, it will appear to others that you are a *lazy project manager*. However, even you will be *pleasantly surprised* to discover that *progress is being made* with almost *no effort* on your part; that *everything is stacked in your favour*. Your customer, your subcontractors and colleagues, your management *respect* you. The company's CEO is calling you and asking you how you do it. And you get used to that *relaxing feeling of comfort* and *full satisfaction*.

You get paid great money, the *money* you think you are *really worth*. You have *enough time for yourself* and *your hobbies. Your family* and friends *fully enjoy* the aura you bring back from work being a *totally happy* and *fulfilled* person.

There are a few basic reasons why I have written this book. The first is that I enjoy writing. I love the feeling of sharing my ideas, making an impact and connecting with you, my reader.

Another reason is that I am compelled to share my knowledge and experience. It is torture for me to keep things inside when I know that this information is invaluable for people embarking on construction projects.

Therefore, the concept of this book came about when I discovered that I wanted to share with you why a building services refurbishment of an existing fully-functional building is completely different to a new-build construction project.

Upgrading a hospital's fully operational heating system in a children's critical-care ward during the winter will stretch you as a project manager, so if you think you will glide through a refurbishment because of your experience in the new-build market, I hate to disappoint you: your client will expose your incompetence and inexperience right at the beginning of the design and installation.

If you are about to manage a building services project and wish to avoid the mistakes and embarrassment which may cost you slipping down a rung or two on the career ladder, then read on.

Industry lacks plain guidance and is full of unreadable and non-practical literature that does not tie up to any structured and coherent approach of managing building services retrofit projects. With this book, and your initiative, we will aim to change that.

How retrofit and refurbishment is different from the new-build construction

Let's get the record straight regarding the differences between new-build and the refurbishment of building services. Here are a few key points:

Costing

➤ Refurbishment is much harder to price accurately. You cannot possibly assess all the variations and things that could go wrong. For example, if the isolation valves will not hold, this will require a whole-system shutdown with disruption to the core business, as the whole system might need draining and refilling.

General management

➤ Your programme will slip dramatically and there is nothing humanly possible you can do about it. Arranging shutdowns will become your top priority. More often than not, there are more stakeholders involved and hoops to jump through than there are fingers on your hands;
➤ In building services, refurbishment design often carries over into the construction phase. As much as we hate this to happen, the existing systems will always surprise you during the installation, causing a re-think of some design elements;
➤ Often, there is not enough space in which to put your new infrastructure, as older buildings do not have plantrooms, high ceilings or risers;
➤ You will often have to work outside of normal working hours;
➤ Your client will have multiple stakeholders: project team, maintenance team, subcontractors and more importantly all-mighty heads of departments occupying the buildings and all with conflicting agendas and egos to satisfy.

Site set up

➤ Access to some areas will be a nightmare, as different departments will have different access keys to the rooms under their control, and no one will be able to find a master key to open them. Often, even security will not be able to help you gain access;
➤ The building will have a fully functional fire system, so your hot or dusty works may trigger alarms. So do not forget to demand that sensors are isolated; otherwise, you can expect to receive a bill for a fire engine turning up unnecessarily;
➤ There will be minimal space allocated for you and your subcontractors' offices;
➤ Existing electrical infrastructure might not be able to support your requirements for welding or temporary boilers. You might struggle to get clean water on site for raw flushing, as the cleaning of water tanks is not always done.

H&S

> ➤ It is so much harder to manage H&S during the refurbishment of existing plantrooms, as you are not in a sole control of the work areas;
> ➤ Existing plantrooms represent a different set of H&S risks and hazards because they were built to the building regulations of their time, which are now obsolete;
> ➤ Making progress with your works in a building full of asbestos-containing materials can prove harder than you think.

Maintenance issues are your problems

> ➤ The maintenance team will be blaming your project for any-and-all issues caused in the area you work in;
> ➤ Cold water, heating, steam, condensate, chilled water (CHW) and sewerage leak repairs will need sorting out before you can crack on with any pipework insulation. This will require lots of organisational effort on your side.

What qualifies as a building services retrofit, what drives the scope of works and the importance of job specifications

So, what qualifies as building services retrofit? Here is the way I see it: a retrofit is a refurbishment of an existing, functioning building; more specifically, the refurbishment of the building's building services and fabric.

Interior and exterior decoration mostly involves building cosmetics and does not qualify as part of a building services refurbishment.

Common sense might tell us that any new system will be more energy efficient than the old one. This is a mistake. The truth is that The Buildings Regulations have advanced so far in terms of comfort and other requirements that new buildings use more energy than those built 40 years ago.

Under the retrofit, you should have a detailed list of works. The scope will grow in depth during the design phase by carrying out validation surveys of the existing systems, developing the specifications, technical submittals, schematics and layout drawings.

See Appendix A for the list of some works which would qualify under building services retrofit and refurbishment.

What drives the scope of works? The importance of the job specification

Apart from the list of works issued by the client, the scope will be very much driven by the condition of the existing systems. To assess the condition of the systems, you need to complete a thorough validation survey (also known sometimes as a dilapidation survey). This is a survey of the existing systems, mostly, to find out what's wrong with them in order that you can set a price for fixing issues in order to make sure they work effectively.

For example, when replacing the pumps, you should also specify:

1. Replacement of faulty isolation valves;
2. If the current system does not have any dosing pots, make sure you add these to your scope of works for every system;
3. If water samples show signs of iron oxide, then make sure this is also specified for the water treatment system.

In your contract with the customer, state that your quote is based on the existing systems being operated and maintained as per current regulations, best practices and industry recommendations. Most importantly, understand the practical application to your project. If you understand, then you can specify to the customer where and how the systems are not maintained as per assumptions and exclusions in your contract. This should let you off the hook for many maintenance issues you might inherit during the project.

Specify that fixing existing maintenance issues, such as leaks, back-feeding from another system, systems crossover and over pressurising, are not part of your contract.

Beware: as soon as you touch the system and do something with it, the whole thing might need to be upgraded to meet current building regulations.

Be aware that electrolytic corrosion may also cause pipes to fill with sludge. Challenging systems commonly found are DHW with galvanised steel pipes joined to copper. Be aware that a packaged water softening and treatment plan will not correct or prevent the problem of electrolytic corrosion.

The older the system, the more problems you will need to solve, so validate the existing systems prior to writing the specifications. If you do not specify these things at the beginning of the project, they will not be priced and, therefore, not completed. It is as simple as that. In the end, and if you're not careful, the refurbished systems may even perform worse than before, at least from a maintenance point of view.

Go the extra mile in finding out more about the maintenance issues in the areas the project takes place. Maintenance records (how lucky you are if you can get those) will help you to define the full extent of the project.

Operation and Maintenance (O&Ms) manuals can reveal the design parameters of the systems, so go through these as well. Use valve charts and schematics in the plantrooms to help get your bearings.

The same applies to the electrical infrastructure. If you are planning for some big electrical panels or chillers, do make sure you survey your current electrical loads so that the infrastructure can take it.

Coordinate your project with the maintenance and capital projects. Do not duplicate something already planned. Get hold of, and fully understand, the capital projects plans for the next year. Works might need to be coordinated with other contractors working for the client.

How important is the job spec?

It is critical for the project's success. You can specify Rolls Royce or Russian Lada specs: both will drive you around in terms of function. The Rolls Royce version will be virtually maintenance free; the Lada version, however, will need constant, unpredictable, reactive maintenance repairs. The Lada version may be cheaper to buy, but it will be very expensive to run.

The same applies to the heating, ventilation and air conditioning (HVAC) systems. If pricing is done by competent and experienced contractors in this field, then the cheaper quote will likely end up being of Lada quality. It is not always the case, however, and more importantly, you have to invest and pay someone who will develop the specifications for the project. This will give you an opportunity to do a competitive tender and compare quotes like-for-like.

This is particularly important for design-and-build contracts where the contractor is taking responsibility for the design and construction of the systems. As with the example of Rolls Royce, you have to specify what to design via a comprehensive scope of works. Otherwise, you cannot compare the proposals for the design and build contractors like-for like.

This is specifically applicable if you are not experienced with refurbishment projects. A word of warning: having a very broad scope is bad, because the design and installation will tend to come down to the lowest common denominator, i.e. price. Having a top-end consultant who never worked previously on any refurbishment projects and simply copies some new-build specifications is just as bad. Top-end consultants will want to earn their money by specifying a Rolls Royce standard of service even if you are happy with a Lada version.

Designers need to make sure that any design complies with current regulations. The lagging has to be as per current building regulations. Valve jackets must be of the appropriate quality.

All building services equipment needs to be as energy efficient as possible. Such equipment includes: Plate Heat Exchangers (PHEs), variable speed drives (VSDs) with pressure control and energy and flow meters of the highest accuracy.

Become very familiar with the manufacturer's installation instructions. For example, one common mistake is not having a long enough pipe for straight runs before and after the water, heat or steam meters and commissioning stations.

Allow for mechanical isolation for maintenance; allow for drain points, vents and safety valves to run to the nearest gullies.

Gullies need to be verified through the RFI (Request for Information) process to make sure that they do not leak. Drain cocks to be allowed in order to connect your new equipment.

Make sure you understand and clarify the proposed welding procedures, as some clients are particularly fussy about welds and do not want any MIG/MAG welding on the job at all, not even on low pressure LTHW systems. Understand and specify the type and thickness of the pipework used (for example, seamless SHED 40). Some clients are touchy regarding the type of pipe and accessories connection: for example, screwed connections might not be permitted for pipes over 32mm, only welded or flanged. The lead time for flanged isolation valves, strainers and non-return valves above 32mm can be weeks. So plan works accordingly.

Pipework spool pieces might be required to come off the imperial-sized flanges onto the new metric system connection. Only ball valves might be allowed on the steam or even in the plantrooms. Steam condensate pipework might need to be stainless steel with no screwed connections as well.

Mechanically, the main cost will be in the equipment, procedures and materials used; for electrical systems, the bulk of the cost is in labour and panels.

If the scope of works is generic, the contractors will come back with mismatching quotes and you will be bewildered why some of them are so expensive. The only way to have a competitive tender is with the designers and contractors working towards a comprehensive scope of works.

19 expensive mistakes to avoid when pricing the job

Pricing the refurbishment of building services is a very, very tough task, and I am almost 100% certain, that you will not get it 100% right. You will either allow for all the eventualities described below and price yourself out of the competition; conversely, your finances and reputation will take a hit should you under price what it takes to complete the project successfully.

Before pricing the job, make sure you fully understand and account for the following things. Even if you want to be very competitive, make sure you specify that you are not doing those things listed below in your offer. Discuss exceptions from your offer with your client who should appreciate that you have fully understood his project. This will reduce the risk of the project's failure in the future. Remember: you design the job on paper; then you build it.

1. Replacing pumps

When you change the pumps for more energy efficient versions, check a couple of things:

- ✓ The power supply may need to change from 3-phase to single phase. This often happens if you decide to use Grundfos Magna3 circulating pumps;
- ✓ Make sure the inlet and the outlet of the pump you install more or less matches the connection of the existing pipes—the inlet pipes and outlet pipes might be in-line or have 90 degrees between them. Doing this means that you can avoid very expensive pipework alterations;
- ✓ When replacing the pump, make sure you match it to the flow rate and the pressure head on the pump curve. Do not allow a salesman from Wilo or Grundfos to oversize the pumps so that they can earn larger commissions for themselves. The best way of sizing the pumps is by measuring the flow rates and working out the pressure from the commissioning stations. If there isn't one, I strongly recommend you install one. The existing systems are often altered (extended or shortened); this means that you cannot trust the flow rates and pressure drops from the original schematic.

2. Isolation valves (IVs) and rotten pipes

Lots of contractors and clients burn themselves by assuming that the existing isolation valves (IVs) are operational. This has three impacts:

1. The replacement cost of IVs and the cost of draining the systems or freezing need to be accounted for;
2. Disruption of the heating, DHW, steam and cold-water services may not be acceptable for the building's users. In other words, you might be requested to re-plan those works for later;
3. As a consequence, this will have a drastic impact on the program, your preliminaries and your profit margin.

To avoid the problem, complete a validation/dilapidation survey prior to signing the contract. Another option is to ask for maintenance records confirming the operation of the IVs.

On systems over 25 years of age, the distribution headers and pipework could be in such a rotten state that it might be impossible to weld a flange or even cut a thread. This will have a very serious impact on your costs and the duration of the project. Pre-empt any of these issues by carrying out a quality validation survey of the existing systems.

Older building may have a mix of systems with imperial and metric pipe sizes, connection types and flanges. If an existing flange is type E, you cannot fit your type D flange isolation valve. Fitting a new valve might involve welding a type-E flange to a spool piece first. All of these are extra costs with a long lead-time for delivery of the type E imperial type flange and requires extra time for installation. When you are later under pressure for completion and switching the heating on, you do not need extra aggravation because of this small oversight.

3.   **Poor water quality**

Making the assumption that LTHW, CWS, DHW water is treated in accordance with current standards could be suicidal. Modern equipment like condensing boilers, plate heat exchangers, pressure-independent valves and AHU coils will clog up in under a year if water is not treated properly. Furthermore, re-doing the works might bankrupt your company. To avoid the problem, take water samples during the validation survey before you sign any contracts. The existing pipework could have all sorts of issues, such as bacteria, sludge, iron oxide or unacceptable PH levels.

Get a specialist to test water in the lab. After sharing the results with the client, you will have two options: either the client's maintenance team will have deal with the problem, or you will be asked to price for these works accordingly. A word of caution: some of the existing systems are already far behind their life expectancy and are the cause of lots of leaks; also, there are still some gravity systems. What this means is that high velocity and chemical flushing normally used in the new-build industry will cause even more leaks to systems that are already aged and troubled. Furthermore, *you* will be liable for damages.

So, use competent water-treatment companies to advise you and to complete the water treatment. Some all-in-one water treatment and filtering equipment is very expensive. However, in the long run, clean pipework and treated water means a longer life expectancy for the equipment, low operational and maintenance costs, and an energy efficient system.

4.   **H&S responsibility**

H&S is your statutory responsibility. You are legally obliged to carry out the duty no matter what. Construction projects specialising in refurbishment will present you with much higher risks than new-build projects. Make sure you understand the job, risks involved and what it takes to complete it safely.

> ➤ Make sure you allow for at least part-time H&S support (a day a week as a minimum). It could be an advisor or H&S manager;
> ➤ You will require some specific personal protective equipment (PPE), such as special gloves, masks, first aider kit, harnesses, fire resistant hi-vests, fire marshal and first aider vests, fans

to cool down some working areas and mats to keep the areas outside of the plantrooms clean. The list of PPE depends on the recommendations made at the risk assessment phase;

➢ Combustible materials will have to be securely stored daily, so you might need a lockable cage. COSHH (Control of Substances Hazardous to Health) items will also need to be kept locked securely;

➢ You will require H&S signs specifying the required PPE and barriers;

➢ Older buildings will have plenty of confined spaces. You might need to send some of your personnel to a confined space course. You will also need to hire a tripod for rescuing people from manholes together with gas meter (sniffer), temporary lighting, kneepads, harnesses, etc. Similarly, the above might apply to restricted spaces;

➢ It is a good practice for your site supervisors to attend the Site Supervisors Safety Training Scheme (SSSTS) or Site Management Safety Training Scheme (SSMTS) courses to understand the full extent of what is required by law and their responsibility if things go wrong. Construction Design and Management (CDM) regulations escalate your costs enormously. As a principal contractor, the list of your responsibilities will be overwhelming and start from H&S, environmental management, security, access, emergency management, and responsibility for the wellbeing of the workforce, visitors, public and even trespassers. The HSE (Health and Safety Executive) might visit your site to check on you;

➢ You will need an Appointed Person (AP) for steam shutdown and an AP for electrical isolation. Those people have to be trained, have enough knowledge and skills to do that job and be allocated enough time to run the safe systems of work, which is almost a full time job;

➢ You also might incur the other costs like DBS (criminal record) checks, compulsory asbestos awareness courses and unproductive time for toolbox talks etc.;

➢ No matter how clinically you arrange the shutdowns, some work will inevitably involve working on live electricity panels, which can only be used as a last resort. Working with live electricity will need appropriate PPE, such as rubber gloves and rubber mats. Trust the advice of your electrical AP.

5. Out of hours work

Allow for plenty of out of hours work. Shutting the systems down or working on electrical or mechanical infrastructure might only be possible outside of normal working hours during the nights and weekends. Bear in mind that quality supervision will be required during these times. This might double your overhead.

In order to run smoothly, all materials, Risk Assessments and Method Statements (RAMS) and labour should be arranged like a military operation. The guys actually doing the work will need to know the exact sequence of works in advance.

6. Builders works

Think through your project, because you will need plenty of builders work to progress with your main packages of work. This will include structural surveys, drilling, building temporary walls, repairs to roofs, walls and doors, fire compartmentation, making good after the removal of existing systems such as radiators including plastering and painting. Equipment like electrical panels, pumps, skids, plate heat exchangers, chillers, boilers, and fans will need to be elevated on plinths in case of floods in the plantroom.

7. Gas works

If you are involved in works associated with gas-supplied equipment, like boilers and CHPs, understand that existing gas infrastructure will need a thorough survey. New boilers need higher pressure, and new equipment often needs to be supported by the upgrading of existing gas pipes, gas meters and gas valves. CHP is an enormous consumer of gas, so make sure you fully understand the cost implication of upgrading gas pipes.

Bear in mind that even some biomass boilers need a tiny gas supply to keep them going. Talking about biomass boilers, bear in mind that they will require quite a large electrical supply as well, so take in to account their electrical and gas-supply requirements.

Biomass boilers and absorption chillers need standby sources of LTHW and CHW. This doubles the capital cost of installation and maintenance.

8. Draining, filling the systems and venting the systems. Freezing the pipework

Some older isolation valves will not hold. You might need to drain some sections of the pipework or a header. Often, you will have to drain the whole system.

- ➤ Check with the estate team which gully you can use, otherwise you might receive a bill for flooding some rooms underneath if the gully and drain pipework is leaky;
- ➤ After completion of your works, you will have to pressure-test the new parts, flush them and fill the system again. Use clean water and do not forget to chemically dose it. Ask the estate team what they are using for chemical dosing;
- ➤ Make sure that water softeners are in place and operational, and factor in the cost or exclude it from your scope;
- ➤ Some systems, such as wet heat recovery or chilled water, will require a proportion of antifreeze, like ethylene glycol, propylene glycol or similar;
- ➤ After filling the system through the pressurisation unit, make sure you vent it through the automatic or manual vents and the radiators. This could be a task in itself and especially if access to rooms is restricted;
- ➤ No doubt you will have to freeze pipes at some stage of the project. If the system is massive and the existing isolation valves do not hold, allow cash for freezing the pipework. Freezing will act like an isolation point allowing you to work on the system without draining the whole building. Freezing will involve shutting the existing system off temporarily to allow pipework to cool down so that freezing with liquid nitrogen can take effect. The pumps will also need to be stopped for the same reason;
- ➤ Arrange wet-systems shutdowns with your customer, as there will be no heating in the building. Furthermore, AHU coils will not have heating, and if the temperature is below 16 °C outside, then you might need to turn them down on the inverters or switch them off for the duration of works.

9. Allow for producing the installation drawings

To cut corners, speed up the process and save cash, many contractors would love to install from the schematics and construction issue drawings. By installation drawing, I mean a drawing developed from the construction issue layout drawing showing the dimensions of the spool pieces, flange to flange dimensions, scalable accessories, etc. This is the drawing used by welders in the workshop so they can weld the spool pieces ready for the pipefitters to bolt them in to place in the plantrooms.

Unless this is something really basic, such as replacing the pump or a fan, specify that installation drawings are compulsory in your contract.

Some contractors will want to install from the schematic and then produce as-built drawings. Do not permit this: you risk huge coordination, structural, space and technical issues rising up and then falling on your shoulders in the middle of any delayed installation. Electricians will connect the electrical equipment and the mechanical contractor may well then want to move the pump to the left to fit pipes.

An approved fully coordinated installation drawing will eliminate the problem of the services running into each other during the installation.

The same goes for builders work. The holes will be drilled, plinths built and structural surveys completed, but all of that work might need to be re-done when the mechanical contractor turns up and says that the location of the plinths and holes is wrong. The mechanical contractor can also simply change his mind.

10. Upgrading pipe sizes

You will often have to allow for an increase in the diameter of pipework. This will often apply to the steam systems. If you are replacing steam/LTHW and steam/DHW calorifiers with plate heat exchangers, then both steam supply and condensate return might need to be upgraded. Any heating/DHW load will dictate the estimated pipe sizes.

To calculate the heat load, allow for the fitting of a strap-on flow meter on the CWS feed for the steam/DHW calorifiers. For the steam/LTHW calorifier, install the heat meter on the LTHW return. Install the meters for a two-week period at minimum. When using the heat meter, make sure that you obtain readings when the temperatures outside are cold so that you can apply some factor to work out the maximum heating load when the temperature is at its lowest. You can hire strap-on flow meters or buy them.

Download the data from the meters and send it to the manufacturer of the plates. They will be able to select new plates with clinical precision based on that data. Do not size the calorifiers based on the manufacturer's data plates of the existing calorifiers, size of the steam pipes feeding it, the size of the steam valves or the size of the pressure reducing valves. If you do, then you will massively oversize the plates, which will cost you more; and because the plates will work on a partial load, this is likely to cause scaling. Another common problem with oversizing the plates is increased wear and tear on the steam control valve, which will be operating at the bottom of its controllability. In turn, and in time, this will erode the valve seat; consequently, the control valve will not be able to shut fully thereby causing high temperature lockout that will then require a manual reset.

As a result you will be replacing the control valves every year due to warranty or design mistakes. The problem is that it is not just the cost of the valve but the added costs of a heating shutdown, labour and the time arranging the above.

There is another big problem with oversizing equipment like steam plate heat exchangers: oversized PHEs need a minimum flow rate through the plate so that it does not lock itself out at high temperatures. For example, if your existing DHW system runs at 1.0 l/s on DHW return and you install an oversized steam/DHW PHE requiring a minimum 1.15 l/s to operate, then your plate will be constantly locking itself out on higher temperatures. Upgrading the DHW circulation pump and installing a bypass might not solve the problem of an oversized PHE.

11. **Limit the liability**

The client might push you to accept the liability for disruption of the business in the occupied building and damages to the building assets. If you do not limit your liability in your contract, it might cost your company. Take appropriate insurance and consult a lawyer if the contract is sizable.

If things go wrong during a replacement of the existing transformer on site with a low loss one, part of the estate could potentially lose power. Although all critical equipment should have a critical power supply and be supported by generators, CHP and Uninterruptable Power Supply (UPS), you really are on thin ice in the event of failure.

Some voltage optimisation (VO) equipment has no bypass switch. When VO becomes faulty, you lose power for whatever it feeds! For smaller estates, this could mean that the whole site goes dark. Use the suggestions in this book to fully think through the impact of things going wrong during your refurbishment project. More importantly, assess the risk by completing a risk register with the estates and facilities team; this will give you a chance of understanding the impact of things going wrong and limit your liability.

12. **Assuming that the electrical infrastructure is in a satisfactory condition to take on your upgrades**

As with mechanical services, survey and validate the existing cabling, switchgear, protective gear and the panels. What follows are just a few examples of how people underprice the installation of electricity consumption reduction measures on site:

1. If there is no earth cable, which may be the case for older installations, you will have to rewire to earth. The existing containment could be completely packed with power and data cabling, but unless you check, then how will you know whether or not you will need to lay a brand new containment? Clipping new cabling to whatever is there (often called a "bow and arrow job") will certainly fail an IEE 17th Edition Wiring Regulations inspection;
2. A Residual Current Device (RCD) will need to be installed in any panels. That is a panel modification. Once you touch the existing panel, it must comply with current regulations. Often, it is cheaper and simpler to get a new panel for both electrical distribution and the BMS Motor Control Centre (MCC);
3. When replacing any basic stuff, like the pump, pressurisation unit or lighting, you will have to test the electrical installation. This involves earth loop impedance, and if you do not have an earth cable, the installation will fail the test. The electrical test can also fail on insulation resistance or continuity test. Any of these are bad news. Rectifying the issues will always involve some serious additional works and associated costs;
4. During the electrical tests, allow a sufficient amount of time for surveys to work out where from the circuit (or panel) is fed from. Also, specify the exact location of the distribution board, as there may not be any existing drawings, labelling or MCB charts available;

5. Some of the existing electrical distribution panels and existing trunking you propose to use for your works will fail the inspection. Before you change anything at the end of a cable, like a pump fed from this panel, realise that the whole installation might have to be upgraded to current regulations. When you open some of the existing panels, you might see bare terminals, no labelling, or even a lack of trunking on 230 V and 430 V cabling and switchgear. You should not be able to poke your finger into any holes that have live terminals.

13. Temporary measures (electrical heaters, oil boilers and temporary pipework)

While pricing any works on heating or DHW systems and business as usual operation, I strongly recommend you allow for some temporary measures to maintain heating and DHW during your works.

In its simplest terms, it could be some oil filled PAT tested electrical heaters, which one can simply plug into the electrical socket. Before you order dozens of them, check the capacity of the existing electrical infrastructure, as you do not want to trip power and lighting for an entire floor.

If you need something more substantial to support the Constant Temperature (CT) circuits feeding the AHUs, radiant panels, FCUs and air curtains, price for the hire of a temporary oil/diesel boiler with a pump set, safety valve, flue and double skinned fuel tank. Most hire companies will be more than happy to come and install one for you using temporary hoses. They will also be able to advise you regarding location (the flue must be at least a few meters away from any windows); they will also advise on fuel delivery accessibility. Ask for some type of call-out maintenance support in case the hired equipment develops a fault.

In a worst-case scenario, there might be no facility to connect a temporary boiler to the existing LTHW or DHW system. What I mean is that the system might not have any isolation points to connect into.

For DHW of a smaller size, say up to 32 kW, you might use an electrical boiler. An electrical boiler will need to come with a buffer vessel to smooth out peak loads. Make sure that you include pricing in the electrical boilers' wiring, as they can, and likely will, consume tens of kilowatts. This power needs to come from somewhere.

14. Commissioning and balancing

OK, let's say that you have installed your new pumps, distribution header or plate heat exchanger: how will you commission and balance the new system?

I bet you forgot my advice and have not taken the flow and pressure reading before you stripped the old bits out. Now, your commissioning company specialists come and ask you, *"What settings would you like your circuits set to?"* While you are pondering on that question, estates and facilities receive complaints that some parts of the building are cold.

No matter what, always measure the flow and pressure for each of the circuits before you strip anything out. More often than not, it also means installing the commissioning stations or meters on existing circuits.

Whilst you are doing that, I strongly advise that you install balancing valves as well.

So, when it's time for the commissioning of modified systems:

a) You know the pressure drop and flow rate it is supposed to be running at;
b) The commissioning company have the tools, such as balancing valves, to balance the circuits of the system.

You either allow for the commissioning sets and balancing valves if they are missing, or you ask your customer to install them for you. Either way, you should discuss with your customer that these have not been priced in the contract. Be ready for it to be an uncomfortable but essential discussion.

Bear in mind that your client will withhold your final application for payment until your contractual works are completed. As commissioning is a part of your contract, you will have to get back to site to commission the systems after your customer installs the double regulating valves on the existing circuits.

If you refuse to return to site, then three things will happen: one, you will have a an unhappy customer who will withhold the final payment; secondly, your company will take a financial hit and a reputational hit; and third, you are certain to get a roasting from your line manager.

15. Resilience

We touched base on resilience before when discussing temporary boilers and electrical heaters. In my experience, your biggest problem with a heating shutdown will be with the AHUs. As buildings have some thermal mass, they can stay warm for hours or even days; however, this is not the case if AHU heating coils heat these areas. If the CT circuit goes off, you are introducing cold fresh air straight into the space.

If the AHU has a re-circulation or heat recover system, then you're in luck. However, blowing 7 °C fresh air with lots of air changes per hour will create a very uncomfortable environment in the space.

In order to work on heating systems in the colder months, you will inevitably have to shut down the heating for the AHU coils. However, it might be possible to tap into (connect to) the existing AHU circuit via the existing draining or venting points. This will allow you to feed the AHUs from the hired temporary boiler. To connect to the draining or venting point of the flow and return, you can use temporary hoses. Do not forget to isolate the flow and return pipes in the plantroom for this to work.

You might also turn the AHUs down via the BMS, or via the inverters manually, by ramping down the frequency.

16. Damages to existing systems and building

Inevitably, your team will damage other installations and existing systems during the installation. For example, you could dent AliClad lagging protection and damage the flat roof if you are working there. AliClad is relatively easy to replace; however, if water seeps under the Aliclad, then it likely that the lagging will be saturated. This means that both lagging and lagging protection will need to be replaced.

If you do install equipment such as chillers, pipework, brackets and accessories on the roof, it is very likely that you will damage the roof's water protection. No matter how well you cover felt or asphalt roofs with plywood and boards, you will damage the waterproof seal. This could be by through swarf or hot rods if you weld there, or bulky and sharp-edged materials like the flanges of heavy pipes and similar.

On a very hot sunny summer day, even your safety boots can leave permanent footprints on a felt or asphalt roof. Leaving any heavy materials will most certainly damage the roof and you will face a consequential claim from the client. The message here is very simple: plan your works, prefabricate off site, cover the areas with plywood for protection, tape over the joints between the adjacent plywood, tidy up daily, and think how your installation will affect someone else's existing work.

✓ Have a workforce that cares

17. BMS dilapidation survey for compliance and operation

Existing BMS systems are another can of worms. All serious projects will involve rejuvenation and optimisation of the BMS. Let's talk about what this means.

You might have some sort of existing system on site, or you can have a combination of different systems with stand-alone control. Depending on how the BMS is functioning, centralising the systems into one common head-end PC might involve the upgrade of the network controllers, additional network wiring and upgrade of the server.

The bottom line is this: to effectively use the BMS system for energy efficiency, maintenance and operation management, you will need remote access to the system. Some BMS systems share the client's IT network.

Most importantly, invest your time and money into validating the existing BMS system. This means sitting with the BMS specialist and going over the whole BMS system, operation of the sensors, actuators, valves, AHU control, chiller control, heating control, monitoring points and the BMS itself. This also means going through the stand-alone systems that are not on the BMS and then pricing for consolidating them all into one system.

This also means talking through the maintenance inspection sheets with a BMS maintenance specialist (and you'll be lucky if can find a maintenance specialist on site) and finding out more about the problematic systems. As a result, you need to understand why the systems are running "in hand" and estimate a price for rectifying the faults.

Rejuvenating the BMS will also involve going through the description of each operation of the system, pricing for having a graphics page for every system, including monitoring. This also means checking the alarms and hard-wired interlocks, like gas safety valves, push buttons, fire alarm points, heat detectors, gas sniffers, smoke removal systems and many more. This will also mean surveying the BMS panels as well, because some of them will need to be replaced with new ones.

It might take weeks for two guys just to evaluate a fully defined scope of works.

After the scope has been clarified, it might take the subcontractor up to three or four weeks just to produce the quote. Normally, this is because BMS installation and panel manufacturing will be subcontracted.

18. AHU flow rates

If your project involves work on AHU and ventilation systems, then a word of caution: as it applies to the water systems, always, always hire a specialist to measure the flow rates and pressure in the systems. Based on current regulations and recommendations, type of area served, volume of the area and occupancy, the consultant can advise you if current ventilation is adequate.

The last thing you want to do is install a VSD on supply fans that already underperform. Over the years, many systems will have been extended and/or modified and you will not find any records. This is why this step is critical. Ask your customer to provide you with a cross-referenced list of AHUs to the floor areas, sizes of the areas in m^2 or m^3 and occupancy type.

To make things worse, you will often notice that some of the rooms have changed use. There will be other situations when someone will have to physically survey the system by following the ducts to understand if the system is performing.

19. Miscellaneous items

I suggest you allow for the following in your contract:

- ✓ Remove waste by subcontracting the delivery and collection of the skips. This is absolutely critical for customer satisfaction: if you have direct account with the waste removal company. This will give you speed and direct control;
- ✓ Storage facilities;
- ✓ Site cabins;
- ✓ Mess facility, drying area, changing room;
- ✓ PAT engineer;
- ✓ Full-time site labourer;
- ✓ Document controller;
- ✓ O&Ms and H&S file specialist contractor;
- ✓ Engineering support;
- ✓ Allow for step-overs, as you have a responsibility to protect your work;
- ✓ Include tarpaulin for protection against the elements, appropriate means of access, such as podiums and platforms, as stepladders might be prohibited on site;
- ✓ Temporary pipework and temporary heat sources, like electric and oil boilers, both of which you will need to provide an electrical supply;
- ✓ Oil / diesel boilers and fuel tanks must be fenced off, spillage kit provided;
- ✓ Be prepared to pay your client for the fire brigade turning up on site when your hot works trigger a fire alarm, overdo toast in your cabin or forget to liaise with the fire system contractor to isolate the smoke or heat sensors in the plantrooms or adjacent corridors.

Contracts: protecting yourself against unreasonable client and making your subcontractors perform successfully

Here is the astonishing truth about contracting.

Most of the contractors in the building services industry, even more in the refurbishment market, do not know what they are doing.

Unless you make specific project requirements very distinct in your contracts with them, it will cost you tons of cash, aggravation, stress, delays, and, ultimately, an unsuccessful project.

Contracts have to be very specific. In effect, they should be covering you financially and commercially from past burns and scalds you may have previously suffered in similar projects with your client and your subcontractors.

The contracts should capture the sales pitch or the promises we are all guilty of making to get the job we want. You can do this by attaching agreed minutes from the tender presentations and all contract pre-award meetings. An enthusiastic sales force and directors will make lots of promises at this stage.

If you do not minute these promises and attach them to the contract, these promises will most definitely be broken. Over time, different people will join the project and it will all boil down to fulfilling contractual obligations and not what was promised.

A project manager, an experienced site manager and building services designers should join the project at the tender stage, and during the pre-award meetings, to select the best subcontractors.

During the pre-award meetings, ask the subcontractors to bring over their project team, including the designers, project manager and site manager.

Ask for three references to make sure that you are not getting a second-class team that lacks positive experience of delivering similar projects previously. Check out the references by calling the referees and asking for balanced feedback. If the team has not completed a similar project, say a hospital, thank them and show them to the door.

There are plenty of companies who have completed projects in a critical environment, such as data centres with cooling systems; unfortunately, this experience is of no use in a hospital environment. If you have a hospital, select the companies who previously refurbished hospital building services.

You are selecting a partner to deliver the project. All the weaknesses you exposed during these few minutes will be multiplied and can become sticking points later in the project.

Never allow the other party to talk uninterrupted for a long period of time during the tender presentation stage and pre-award meeting. They will talk themselves into the project. Have an agenda and questions. Chair the meeting and do not be afraid to cut in if you are not getting answers to your questions or getting no value out of the talk.

Review the items discussed earlier in this book regarding the pitfalls and commonly made mistakes and include relevant sections limiting your liability to the client or passing the issues onto your subcontractors. Clearly, and in great detail, specify what you have allowed for and list in great detail what is outside of the contract.

Avoid general and ever-boring phrases commonly found in the contracts, such as install and commission the system. The ambiguity will cost you lost momentum, pressure and cash as well. A great contract will give you teeth to bite subcontractors or defend yourself against an aggressive client.

A good contract is a document telling you how the job will be completed. Highlight the importance of coordination between the subcontractors and the parties involved. Highlight duty of care as well. Expand in detail on both of the items listed above.

Invest a lot into producing a quality scope of works, design, drawings and specifications. If you have any schedules, sketches, notes and schematics produced during the tender or during the evaluation surveys, make sure that they are all included. Never copy and paste extracts from a subcontractor's quote into your contract; otherwise, you will end up with a reduced scope of works and you will end up working on the terms of the subcontractor. Often, to leverage the job through the senior management of the subcontractor, you must have teeth in your contract.

Do not start working or accept any payments from your client without a signed contract and a purchase order for the value of the project. If you really want to get started, you can work based on a letter of intent. The letter of intent has to state specifically the works to be completed and, more importantly, the value of the works. The contracts need to be signed even for the letter of intent, though.

To learn from your last mistakes, review the variations and site instructions you issued to your designers, supplier sand subcontractors on the last couple of projects. This review process will reveal the grey areas which your previous contracts did not cover and which your subcontractors exploited to their advantage. Make sure that you revise your contracts to cover those grey areas by offloading the responsibility and risks to the companies you are about to hire.

You can use up every weapon available to influence the non-performing contractor in the world including escalation, micro-management, reasoning and motivation, appealing to the good in them, bribery, even threats, and all to no avail. If your contract is weak, the contractor will keep referring back to it and using it as an excuse for non-performance.

So, gear up for it and have Dracula's fangs in your contract ready to bite as soon as things get out of control.

Until the contractor realisers that his bad performance, sloppiness and incompetence will cost him much more than the contract value and that he is on the clock, he will not change.

From experience, and irrespective of the type of standard contracts your organisation uses (NEC, ICT or a bespoke one), here are some key clauses to offload onto your subcontractors:

1. The following documents to be updated and issued weekly to the principle contractor (principle contractor will produce these documents at subcontractor's cost if these documents are not issued):
   a) Procurement schedule with the delivery dates for equipment and long delivery items;
   b) Two-week look-ahead with the labour assigned for each task on a daily basis;
   c) Register of drawings and specifications listing all the drawings, their status and the dates of release for each of them;
   d) Schedule of Risk Assessments and Method Statements (RAMS) listing all the RAMS required to complete the job with the date of issue;
   e) Schedule of Requests for Information (RFIs) listing all RFIs and outstanding items;
   f) Register of technical submittals with the date assigned for each technical submittal required to complete the project;
   g) H&S inspection report including near misses, observations, weekly toolbox talks registers and the weekly inspection registers for harnesses, means of access and scaffolds;
   h) Program with a drop line listing the items stopping the progress of the subcontractor's works.
2. Have a clause confirming that you can replace the contractors' management team if they are consistently underperforming;
3. Contractors will need your confirmation in writing from you before reducing labour on site;
4. Principle contractor has a right to veto the selection of sub-subcontractors;
5. Liquidated damages apply after each missed milestone (make sure you have milestones in the contractual program)—this will allow you to penalise undeforming contractors immediately, which will help galvanise the minds of their management team;
6. A reasonable amount of overtime should be allowed in the contract to carry out works during shutdowns and outside working hours;
7. That subcontractors are to coordinate their works and commissioning with all the other contractors and the building end users;
8. Have a duty of care to the principle contractor, end users, maintenance team and other contractors;
9. Fully coordinate their design drawings with the other trades and design teams;
10. The design consultancy is to develop specifications for LTHW, CHW, DHW, DCW, ventilation, lagging, steam and condensate, valves and accessories, pipework, ductwork, BMS, electrical services and meters; you will also require individual specifications for the equipment starting from chillers, AHUs, PHEs, pumps, fans, electrical panels, BMS panels, cables, lighting and electrical installation. This is to protect yourself and the project;
11. The design consultancy are also to produce fully coordinated drawings showing all the mechanical, electrical, BMS systems including the trays, panels, plinths, holes, builders works and location of the equipment;
12. All associated builders work are provided by subcontractors;
13. Each site operative and management team to provide no more than six-months old criminal record check certificate (for people having access to children it has to be an enhanced DBS certificate);
14. Subcontractor's project manager to allow for a weekly site walk around with the principle contractor;
15. If rubbish is not removed within 24 hours from a notice, principle contractor reserves the right to remove the rubbish on behalf of the subcontractor and back-charge the associated cost from the next application for payment;
16. Subcontractor to have two certified first aiders and fire marshals so that one can cover the other if one of them is on annual leave or sick leave;

17. Each person from the subcontractor's organisation entering the site to have an asbestos awareness course (this is a legal requirement for buildings built before 2000 anyway);
18. Specialist engineers to be hired to commission the expensive pieces of kit, like plate heat exchangers, pumps, pressurisation units, meters, aM&T, boilers, CHPs, BMS and state-of-the-art water treatment plant;
19. Contractors have to allow for specialist contractors to flush, balance, chlorinate DHW and DCW and commission the mechanical systems;
20. Disposal of flushing and chlorination chemicals after flushing to be done by subcontractors in accordance with environmental law;
21. Structural surveys of the road and the drop off area for craneage to make sure the area will take the load—this will include sample drilling of the landing area and lifting area, which is very expensive and disruptive to your client;
22. Structural surveys for lifting blocks before making holes in the walls and floors;
23. All plant movement to do with associated works are by subcontractors;
24. Provision of power for welding machines, flushing, temporary boilers and anything is in the subcontractor's package;
25. Provision of water for flushing is by subcontractor;
26. All site, commissioning and project managers including supervisors and foremen to have Site Safety Management Training Scheme (SSMTS) certificates;
27. Subcontractors to have fully trained and competent Appointed Person (AP) for electrical and mechanical isolations;
28. PASMA tickets for each operative using a mobile tower;
29. For thermal insulation use the following as good practice:
    ✓ Carry out a thermal imaging survey before and after the pipework insulation
    ✓ 100 mm of mineral wool for steam/LTHW/condensate pipes over 80mm;
    ✓ 50-80mm of mineral wool for the steam/LTHW/condensate pipes below 80mm;
    ✓ For the areas outside exposed to rain, mineral wool to have both VentureClad and then 0.8mm AliClad protection on top;
    ✓ The brackets should have calcium or wooden blocks to insulate the brackets from the pipes—you do not want metal rods conducting heat;
    ✓ Step overs to be a part of the contract to protect the lagging.
30. Include the following section into the contract with subcontractors that carry out welding:
    ✓ Welding procedures to be written by the subcontractor and approved by the principle contractor prior the works
    ✓ Welds are to be inspected by the certified and competent welding inspector who needs to be agreed between the subcontractor and the principle contractor
    ✓ 100% of the welds to have NDT. This could be a mix of Magnetic Particle Inspection (MPI) and ultrasound inspection
    ✓ 10% of all the welds to be X-rayed. If faulty welds are found during the X-ray, another 10% of the welds made by that particular welder are to be inspected. If more fail, 100% of the welds made by this welder will have to be inspected
    ✓ A welding map to be provided and to include every single new weld on the project. This map should clearly identify by which welder each weld was made
    ✓ For the steam system the welds to be inspected against the welding procedure.
31. Subcontractors have thirty days to rectify any snags;
32. Adequate time to be allowed in the contracts for witnessing of the systems to the principle contractor and client.

Design and build contract

Design and build contracts are popular. In the current economic environment, everything, including maintenance is outsourced. By offloading the contracts, the organisational financial management reduces overheads and directly employed office staff. The downside to this is losing site knowledge. In our particular case, this is knowledge critical to getting value for money, completing the project and delivering quality work.

For risk-averse organisations, design and build (D&B) contracts is a way forward. D&B means that the organisation contracts a construction company that will take on both design and installation responsibilities. With D&B you will avoid the never-ending battle between designers and contractors.

Contractors will be blaming consultants for designing a project without any consideration for installation methods, commissionability and maintainability.

When things do not work during commissioning, consultants will defend themselves and list dozens of examples where the system was not installed to their design.

Every single design change will be picked up by the subcontractors and then followed by a request for a variation—some contractors hope to make money on variations like this. Then, at a later stage, the designer runs out of money budgeted for the project and stops attending the site to validate the installations. This happens unbelievably often. You will be stuck in the middle of this ugliness and suffer the frustration of being unable to move things forward.

For client organisations, if you are undertaking the project on an existing estate, I suggest you go down the D&B contract route and competitively select the principal contractor to design and install the project for you. All you are left to do is to manage the principal contractor. Every project team of the client wants peace of mind.

Dilapidation surveys

Dilapidation surveys completed prior to the design stage are paramount for the success of the building services retrofit.

The design and installation will be very easy if you invest in comprehensive dilapidation surveys.

Record the details from the pumps' motor plates and fans, including the manufacturer, and schedule them as a document. Wet systems normally have pressure gauges before and after the pump. Record those readings.

Also measure the current drawn by the motors. For the ventilation systems, get a specialist to record the airflows and system pressure.

These invaluable data will give you a head start in selecting direct replacements for the fans and pumps. It will also give you a figure a commissioning specialist can use to balance the systems using the regulating valves and dampers. If you do not have the flow rates and pressures from the original system, how can you make sure that the new system will achieve the requirements of the existing systems, such as required heat load?

For the pumps, for example, you can obtain the pump curve of the old pump based on the manufacturer and the flow rate. System or circuit pressure drop will give you a second figure. Based on that curve, you can work out the flow rate.

Based on the flow rate and temperature difference between the flow and return, you will be able to calculate the maximum heat load, which is one way of selecting the pumps, heating generating equipment, plate heat exchangers and pipe sizes.

Here are some of the critical points to think about:

- ✓ The best way to work out the flow rate is by using commissioning stations, even if you need to install a new one into the existing system;
- ✓ The most accurate way of selecting the DHW and LTHW plate heat exchangers is by installing meters. You can hire strap-on water meters; and when you know the temperature difference between the cold-water feed and DHW flow, you can work out the maximum DHW load for the system. You size the plate heat exchanger or DHW boiler based on that calculation. The good thing about DHW is that logging flow rates will work at any time of the year;
- ✓ It is also wise to install another meter on DHW return. This will provide essential data to make sure that there is enough circulation through the equipment to function without tripping high temperature at low or zero demand;
- ✓ For the LTHW systems, install a heat meter on the LTHW return. Record both flow in l/s and heat in kWh. Obviously, you need to try to log the maximum heat consumption during

the coldest spell of weather. This could be tricky, so apply the load factor to work out the maximum load;

✓ Always measure the natural gas pressures at the point of use prior to replacing any of the heating or DHW generating equipment. New equipment requires a higher pressure drop. This can only mean upgrading the gas meter, uprating the size of the gas pipe and gas valve— these are associated with huge costs;

✓ You must take DHW, DCW, LTHW and CHW water samples, send them to lab and obtain confirmation certificates regarding any existing issues, as they are not for you to sort out;

✓ Dilapidation reports for the electrical and BMS panels will also highlight incompliance issues which you can pass onto your customer or include in your price to sort out later;

✓ Check operation of the accessories and equipment and record any problems;

✓ If you are making any hydraulic changes with the system, make sure that you check the operation of the double regulating (balancing) valves. If there are none, then make sure you price for them to be installed to balance the circuits effectively to the original setting.

Get an independent design specialist to review the preliminary issue drawings from your designers. This chap will compile a report with comments and questions. If the design is bad, he will condemn it as not fit for purpose and will bring up any risks associated with it.

Crucial step. How to select star subcontractors and sub-subcontractors

The key lesson of this chapter is that the refurbishment of existing building services is a niche construction market. When I say niche, I mean the players in the market are used to working on existing operational systems in a fully operational building and have successfully overcome their associated obstacles and challenges many times in the past.

Interestingly, the specialist companies working on specific systems are normally small-sized companies with the typical problems of small-sized companies: they all experience cash flow problems. For example, a £300k order over a 2-months program and with a 60-day invoice payment plan will likely send them belly up. They cannot support labour and materials for too long without being paid. To work with them, you will often have to adopt a more flexible payment plan and give them work in smaller packets.

As a client or principal contractor, the biggest mistake you can ever make is to give a job to a typical new-build mechanical or electrical subcontractor. In fact, this would be a recipe for disaster, both financially and program-wise, and will cause serious trouble for the client and building users.

The rule is basic: no positive experience in refurbishment of building services, then no job.

A new-build contractor is used to coming to site with the building users already relocated elsewhere, everything set up and fenced off as a normal construction site, restricting access to everyone, including the maintenance team, stripping everything, installing brand new systems, commissioning and then handing over.

The problem is that most clients do not want lost services and need the building to remain functional throughout the project.

The most common method of contracting the project in refurbishment is hiring a specialist consultancy to design the project for you on paper. This will involve producing the schematics, calculations, layout drawings, equipment schedules and specifications.

Once the design is completed, you issue it to three-to-five specialist contractors to obtain quotes. In effect, you are undertaking a competitive tender. One of the best ways of finding a company specialising in building services refurbishment is to ask your client to recommend the contractors who are already involved in refurbishing wet systems, lighting and power, BMS and ventilation systems on site.

These companies has proven themselves to the client already, have a relationship, know the site intuitively and are used to doing small projects in a functioning building. The alternative is using similar contractors from your past experience or by recommendations.

Now, ideally, clients want a principal contractor to manage those smaller subcontractors; alternately, the client will need to employ its own project-management team to deal with permits, H&S, coordination, site set-up, security and commercial issues to name but a few. In both options the cost will go up. In the case of the principal contractor taking on the job, the risk of the project's failure will be reduced. This is because the principal contractor will be responsible for the delivery of the project and will be responsible for compliance with construction law.

Ultimately, it means that it is the principal contractor's problem to manage the risks of their subcontractors going bust, managing health and safety (H&S) and delivering the project on time as per contract. In practical terms, you as a client have off-loaded the burden to the principal contractor. This is even truer if the principal contractor takes on the design responsibility or if this is a design-and-build project.

Cascade your quality and H&S systems throughout the supply chain an reinforce it in the contracts

When drafting the contract with subcontractors, cascade your quality systems throughout the chain. If your company is well-respected brand, and specifically if it is a premium brand, enforce your rules on your subcontractors. You should have aspects of ethical, environmental, anti-bullying, equal opportunities, preferred supplier and H&S polices in your contract.

You must have zero tolerance to bullying, discrimination, poor H&S practices and environmental carelessness on your project. This applies to your subcontractors and their subcontractors. In other words it applies to the whole supply chain, including the manufacturers, distributers and logistics companies. Embed your standards and set high expectations for all of your partners.

Your contract should stipulate the enforcement of the use of the permit-to-work system: daily hot works permits, electrical and mechanical isolation permits, permits to work at height and from stepladders, and permits for confined spaces. You should request weekly H&S and environmental inspection reports from the subcontractors; employ a near-miss system that encourages people to report what is unsafe; request that subcontractors provide weekly toolbox talk signed registers; request that subcontractors' H&S managers or directors attend compulsory monthly H&S walkarounds. Implement a yellow/red card system, i.e. operatives receive a yellow card for minor H&S misconduct and a red card for something more serious and results in permanent removal from site.

Set up an attendance management system (access cards or using a basic sign-in/sign-out register). Request that companies train their operatives, supervisors and managers on improving their competencies. Set high standards for accepted competencies in forms of adequate and specific job experience, CSCS cards, tickets for working at height, etc. All supervisors must have SSSTS, fire marshal and first aid certificates; all managers must have SSMTS.

Enforce a compulsory asbestos awareness course and criminal-record check for every operative, supervisor and manager on site. List the site rules in the contract, as delivering the site rules during the site induction will not be enough.

Cascade your quality and H&S systems throughout the project via the contract, otherwise you will hear later that we have not allowed for all of that in your preliminaries. If you get those updates, managing the project and updating your client will be much easier. If you have these conditions written in your contract, they will be in breach of contract if they do not provide you with all this information.

Make yourself clear; set the requirements from the start.

Project program. To succeed, account for failures

My clients always ask me if the program deadline will be met. For some reason they are anxious about it. It is very important to them, as they think that meeting the project deadline is the only criterion of a successful project. Let's get into the masterful programming of works, then, so our clients are happy.

In order to tie your subcontractors to some sort of duration for the project, you need to create a project program. Here is a word of caution about programs: I have yet to see a program that met the completion date. All projects are delivered behind schedule.

This, however, does not mean that once the last day on the program is finished, your subcontractors breached the contract; they will have some legitimate reasons why the completion was delayed—although some delays will indeed be caused by a contractor's performance on a particular package.

To create a more-or-less accurate program, you need a very good understanding of the site conditions, design risks, the performance of existing systems, historic performance of the subcontractors on a similar project and the quality of supervision and management. Past lessons from similar projects have to be accounted for otherwise your program will not be worth the paper it is written on.

Here is another important lesson: programs are not recoverable. What does this mean in practical terms? If you've slipped behind on your dates, you will never catch up. The project will be delivered behind schedule. No matter how much labour you throw at it, and no matter how many shifts you work, you will never accelerate the recovery of the program. No matter what you do, the program will always slip further behind the completion date.

Instead of focusing on the past and getting frustrated with your suppliers, contractors and management team, take my advice: focus on the task at hand. Focus on immediate showstoppers, the problems your subcontractors are facing, the daily issues. Understand the cause of the delay and act on it with all your wisdom and might. This is what the best project managers do: they find out what the problem is and go straight to its heart.

Outstanding project managers dare uncomfortable conversations, pick up the phone, talk to subcontractors, suppliers, logistics companies, designers and supervisors on site. They escalate problems to directors; they go to the client for help using their relentless resourcefulness to find the solution that solves the problem.

If they are faced with obstacles, they will swallow their pride to get the issue resolved. They focus on the result and drive the program. Soon, people around them understand that this chap is not going away until he gets what he wants and start helping.

They ask questions regarding what could be done, agree an answer and implement it immediately. They have an unrelenting momentum, charge the parties involved with their unstoppable energy and make sure that the problem is resolved as a result. You need to become such a project manager.

The experienced supervisors of your subcontractors should drive the program that you are creating. This means that each package of works should have its own program. Later, you collect the programs from each of those packages, look at coordination issues, sequence the package of works and, based on that, create a master project program including design, installation, commissioning, training and O&Ms.

Each program for the package should cover 100% of this package's scope, be broken down into areas and contain a labour forecast. The works should be broken down for, say:

- ✓ Dilapidation survey;
- ✓ Preliminary design;
- ✓ Technical submittals approved;
- ✓ Design approval;
- ✓ Installation drawings;
- ✓ Mobilisation;
- ✓ First fix;
- ✓ Second fix;
- ✓ Power on;
- ✓ Pressure testing, flushing, balancing;
- ✓ Commissioning;
- ✓ Snagging;
- ✓ De-snagging;
- ✓ Operation;
- ✓ Training
- ✓ O&Ms.

Plan heating and DHW works between April and October. This means your design for a multimillion-pound project should start in June the year before. During the dilapidation survey, you will be able to test the valves to examine if they are holding, survey the areas being fed, measure the flow rates and pressure drops on branches, and measure the flow rates on the AHUs. A dilapidation survey takes weeks and weeks for large estates.

Once the validation is complete, design can commence in the autumn. It will start with the schematics, technical submittals for all of the pipework, ductwork, accessories, schedules of equipment with details and manufacturer's information. Job specifications will have to be developed from the initial scope or specification and then go into more depth. All of these will be debated in design meetings for months. At a later stage, the layout drawings and coordinated drawings could be produced. After that, the installation drawings will need to be completed.

By the end of the winter, all the design must be fully completed. For the BMS this includes BMS points list, description of operation, wiring diagrams, valve schedule and schedule of meters.

The critical step is that the subcontractors need to be selected, orders placed and contracts signed.

By the end of February, all the long-lead equipment, such as plate heat exchangers, boilers, transformers, voltage optimisation units and chillers, need to be ordered. For CHP and biomass boilers, delivery could be up to 6 months or more depending on your design, so order these even earlier. Procurement should be one of the driving forces of the project.

Aim for the site set-up, office set-up, storage set-up and mobilisation for the beginning of March. If you use cabins, it will take weeks to get power, fire alarms, telephone lines and broadband. Your mess room will need a cold water supply and sewerage as well. The drying room/locker room allows wet clothes to dry and provides a place for your workforce to change clothes.

This is also an ideal time for the installation sub-subcontractors to familiarise themselves with the works and plantrooms, order the materials and accessories.

If you have completed your validation survey and recorded the findings, you will know which systems can be worked on during the heating season, which is 1st of October till the end of April. This could be standby calorifiers and plate heat exchangers, standby pumps and control valves, all of which in this example could be easily isolated. So plan those smaller things during the heating season to provide you with a continuity of work.

Tie up with the maintenance team to line up your works when routine shutdowns of the AHUs, CHW, heating and DHW systems occur. Highlight those works as critical and make sure you have every single nut and bolt on site a week before you start, as things do go wrong.

Do your CHW works during the heating season. Comfort cooling should be off, however this may not always be possible; for example, cooling for the data and server rooms, and critical cooling for specialised equipment may be running all year around. So liaise with the client's on-site maintenance team for the routine maintenance shutdown schedule.

If you are not experienced with completing similar projects, then triple your estimates
for the durations of time you calculated on your program of works.

In retrofit, lots of things will not go to plan, so plan for that as well. Do not forget that your client, your designers, your suppliers and your subcontractors will all miss dates. At busy times you will have a bottleneck in your progress, so account for that as well.

Save yourself embarrassment: build huge buffers into the program and break it down as far as possible.

PART 2

Pre-construction

Have fun being the principal contractor

Being a principal contractor is totally different to being a subcontractor, designer or consultant. It gives you power to run the job the way you want, schedule the meetings you want, where you want and whenever you want. It gives you powers to command your direction, demand explanations and stop works on site if there is a need to.

With power, of course, comes responsibility.

Have fun being the principal contractor!

Being a principal contractor is a role requiring specific skills. The most important skills are:

1. Ability to understand where the project is. This is done through a combination of walkarounds, being on the project from the beginning, having the technical capabilities to comment and approve the design, having a very good understanding of the installation practices, workmanship and being able to validate commissioning. You need to have a professional opinion regarding where each package is in terms of the performance, targets and difficulties faced.
2. Have strong leadership skills. You need to be able to direct people and works with confidence. Opinions will vary significantly, but at the end of the day, and especially when everyone starts panicking, it is expected of you to give a clear and believable direction.
3. Ability to select a great team. Have a good eye for great people, but you also need the ability of let go of people who do not perform. You need to have a project brotherhood and good spirit, which should include your client.
4. You need to be a good communicator and listener in order to understand what drives and motivates your team members.
5. You need to have strong interpersonal skills to understand the types of personalities you are dealing with and how to change your language and actions accordingly to get the project moving.
6. You need to be firm and tough to withstand the pressure from the customer and subcontractors. Your psychology and mentality must be very strong. By the way, this comes with lots of positive experience.
7. You need to be driven to get the project moving. The energy coming from you should be infectious and charging people to action, overcome their barriers and work in a team leaving personal differences aside.

The principal contractor has enormous powers, but you need to harness those powers and use them to channel the project in a progressive direction. Being a principal contractor is fun. You have a right to ask almost any questions, make requests, stop works, request all sorts of paperwork and help. Expect the prompt responses from subcontractors fast and without reminders.

Do not shy away from that power; however, do not overuse it. At the end of the day, it is the designers and contractors who deliver the project. You take all the glory when it is finished, but you get all the aggravation and stress when it is failing.

Prior to the commencement of works on site, make sure that your customer provides you with a copy of the signed F10 form from the CDMc (Construction Design and Management coordinator) who will be appointed by your client.

Make sure that the designers produce design risk assessments.

As a principal contractor you have a legal responsibility to produce the following:

I. H&S plan;
II. Environmental plan;
III. Fire and emergency evacuation procedure.

It takes time to develop the above-mentioned documents; however, there are plenty of samples available to follow and help you out. If you are unsure where to get the samples or want some practice, I suggest undertaking Site Safety Management Training Scheme (SSMTS) training, as you will be provided with a pack of literature to refer to when in need.

In effect, you design your site on paper prior to starting any physical work. For the best effect, organise a site set-up workshop with your team.

For fire and emergency, organise a workshop with your client's fire officers to see how you can tag onto the existing procedures in place.

For a quality environmental plan for complex projects, you might consider hiring a specialist consultant to help you out.

After getting great subcontractors and suppliers, use your wits and influence to get the following structure and support for your project team:

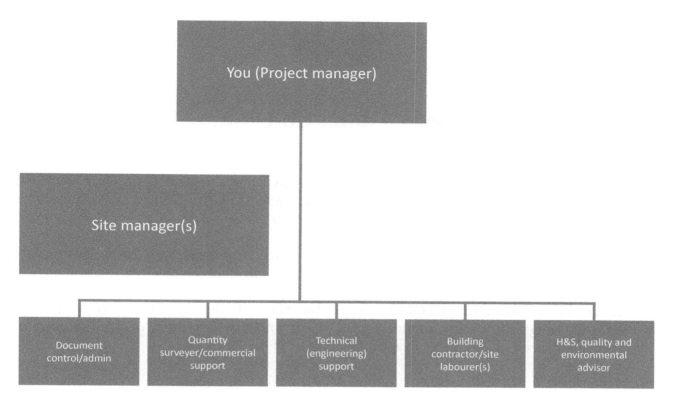

If you fail to get quality support, then your project will struggle or fail.

Fight with all your might and wits to get the best people in your organisation working for you on your project, not on a project of your colleague(s). Threaten to resign if you have to. Remember the old construction-industry saying: you are only as good as your last project.

H&S management – don't find yourself in prison

The more I work, the more I realise that a safe site is an indication of a successful project. H&S is an integral part of the project's management. It is like playing football in a different position in that it gives you an invaluable perspective on the way the project is managed.

The correlations and dependencies between H&S and the project's progress are still a mystery to me. I can only take it from the experience that the cleaner the site, the more compliant the site will be with H&S requirements and the smoother the job runs. I also think that regular walkarounds will help you understand first-hand where problems may occur; talking to the supervisors and experienced workers will also give you some insight as to what and where the real problems are. This critical intelligence will give you knowledge to apply pinpoint focus on the contractors.

H&S is not an option: it is compulsory and enforced by law. Weekly, project and construction managers lose their jobs because of H&S breaches and/or accidents.

H&S tends to be neglected, remaining off the radar until an accident or dangerous occurrence happens. Then, when it happens, the world turns upside down: your best mates, lovable boss and adorable client turn their backs on you. H&S issues are escalated to directors and CEOs, and no one wants to get on the caught-being-bad radar and being associated with it.

> *Understand and prepare for it mentally: accidents will happen on your projects and you will be made responsible for them. Have safe systems of work to protect yourself.*

When things go wrong investigations are launched, explanations requested, H&S audits are carried out, the guilty are punished. The reputations of the most reputable contractors and principal contractors are shattered by serious accidents. It does not just affect next year's profit: you can get a criminal record or even go to prison.

Here's the thing: the Health and Safety Executive (HSE) *always* confirms that it was the fault of the principal contractor. You could have some imbecile on drugs doing something stupid and unthinkable on your site. The thing is that, partially, it is your fault as principal contractor, because you allowed him to work on your site. Therefore, you are one of the prosecuted or penalised. Here, the buck really does stop with you.

H&S needs to be set up by the project or site manager from the beginning. A safe site does not just happen. Be very clear about the H&S goals and the rules. Start by evaluating the contractors' H&S track records before appointing them (type the contractor's name into the HSE website for a start).

Cascade your H&S quality system on your subcontractors via the contract. This system should contain the forms with the permits, isolations, monthly reports, toolbox talks, guidance, near-miss cards, competency requirements and, possibly, even risk assessments and method statements.

Set up the H&S file electronically, but have hard copies also. Create a site induction document. It has to be very specific for the project. Create the site rules. Do not be afraid to have sections on compulsory cleaning and sweeping of the working place: a clean site is a safe site. Enforce weekly toolbox talks on certain days of the week. You can even ban stepladders.

Specify competency requirements for trades such as welders, welding inspectors, pipe fitters, scaffolders, electricians, commissioning and flushing specialists and BMS engineers. There should be no cowboys on your site. Clarify that only the appointed person (AP) can see to steam isolations and electrical isolations. Throw people off site during the inductions if you have reasons to believe that they will not obey the rules or are unqualified for the job.

Set up a permit system on site. No work to be carried out without the permit. The permits should be monthly for general repetitive and less dangerous work. Daily permits are essential for hot works like welding, grinding and chopping. A daily permit is also required for the electrical isolations, work at height and confined-space work.

Buy two key cabinets: one for the keys and chains for the isolation valves and the other for electrical isolations. Buy various devices to lock the circuit breakers so no one can switch them on or off apart for the AP (Appointed Person).

APs should run the permit system for isolations. They will be locking the systems and issuing the permits to work on those specific systems. Once they are satisfied with the completed work, they will unlock the devices so the systems can be used again.

Hot works will be required for any works involving sparks, flames, burning, welding, smoke and even dust. The permit must have a check item on the isolation of fire alarms. The existing smoke sensors, and sometimes sensors in the rooms next door, will have to be isolated during the working day. Liaise with the fire contractor so that they can isolate the alarms daily at the fire panels.

Bear in mind that dust can also trigger some fire alarm sensors, so ask on the permit for these sensors to be isolated for the duration of works or daily if the work is to take longer than a day. The estate's maintenance team normally outsources fire alarm systems maintenance to a specialist contractor, so you will need to deal with maintenance to get the isolation enabled. As these are recorded on the maintenance team's system, the specialist contractor will be issued a job number and will do the task for you. You are not normally expected to pick up the cost for this.

Welding areas should have a welding booth or at least a tent and be screened by a welding screen to protect the eyes of the guys working around the area. The welders must be coded, which means that they have passed the necessary exams for the type of welding they are doing. If they have a ticket for TIG or MIG/MAG only, then they cannot do arc welding. Check their competency against the welding spec before or during the site induction.

Ideally, the welding areas need to be ventilated adequately to remove any smoke. Welders will have to wear flame-retardant hi-vis vests and overalls. They should wear leather gauntlets to protect their hands and wear a welding mask. Inside the welding area, they do not need a safety helmet. Ear protection might be required.

Welding areas should be equipped with a fire extinguisher. All gas bottles, such as oxygen and propane, must be moved to lockable cages at the end of the day. Fire watch is essential for at least one hour after the completion of hot works. If required, the floor of the welding area should be protected with a sheet of metal to protect against fire and possible damage by stepping on swarf. Face masks and goggles must be worn when grinding.

Electrical equipment, including welding leads, welding sets, grinders and others, should be PAT tested every 3 months.

PAT (Portable Appliance Test) is a must requirement for the project: you do not want employees bringing their own electrically powered tools on site without PAT testing. Either enforce it, or better yet, hire a PAT electrician once a day every three months to test all the electrical appliances on site, including office equipment. This will include 110V extension leads, which are easily damaged.

Recruit a competent and preferably independent H&S advisor to come on site twice a week to help you set-up and run the H&S system. Another important reason for having a second pair of experienced eyes checking your site is because you can become blind to certain risky items and take them for granted. It happens all the time: someone new comes on to the site and finds new risks that you have overlooked or simply fell off the radar.

Be ready for multiple risks introduced by existing sites. The older the site, the more likely you will face challenging problems. Here are a few potential risks:

- Buildings built before the 1990s can be full of asbestos and can often found in lagging, ceiling tiles, electrical panels, walls and floor and/or ceiling tiles. Familiarise yourself with The Asbestos Register (and good luck if the documents make full sense). Prior to starting any works, drive your customer to carry out demolition surveys in the areas where you will work. If found positive, get the customer to remove all the asbestos. Do not overlook the old electrical panels, as they may have asbestos containing materials (ACMs) in the fuse carriers.
- Steam, condensate and LTHW isolation valves that do not hold. There are often no safe methods of testing them and the existing system may lack bleed valves or drain cocks. The best way to test is to allow the system to cool below 35 °C prior to draining or connecting to the system.
- The existing steam, condensate, LTHW and DHW systems are rarely fully thermally insulated. As a rule, bits and pieces of lagging will be missing, including the jackets and covers. Electrical, BMS and mechanical installers are in direct danger of touching those hot surfaces during the installations unless you lag those at least temporary with duct blankets or similar.
- Confined spaces. There will be plenty of those in buildings built over 40 years ago. Confined space training will be essential for personnel entering these areas; emergency procedures will also have to be put in place. A permit system is a required by law. This will slow down your progress substantially; however, you can't cut corners with H&S.
- Hot and humid environment. Be prepared to face plantrooms with temperatures over 40 °C. Think about temporary cooling, such as portable axial fans, wearing leather gauntlets, long sleeve hi-vis vests, lagging pipework prior to the beginning of major works, and do provide water with frequent breaks.
- Weil's disease. If you have rats on site, this is another major risk. Have plenty of medical plasters on site to cover cuts, and provide welfare facilities with soap and hot water. Encourage employees to regularly wash their hands with soap.
- Existing sites will introduce an enormous number of slips and trips. Consider step overs, barriers, and the installation of chequer plates to cover gaps and holes.

- Falling from height is another major risk on existing sites. Splash out on collective measures, such as barriers on the roof. Install them permanently so that FM personnel can continue to make use of them when your project ends.

In addition to the items above, noise can be a serious risk. There are many plantrooms with compressors, big pumps and large AHUs that combine to produce noise above 80 dB. Constant exposure to this kind of noise will have a deteriorating effect on your hearing over time. Special attention should be paid to diamond drilling, which you will need to use to bore holes in the walls and floors for pipework and ductwork. Adequate signs requesting that the workforce must wear hearing protection in a particular area will have to be installed. Moreover, provide earplug dispensers next to the signs so people can use them before entering the area concerned.

Upgrading the BMS will inevitably involve working in the control sections of the panels with 230V. By law, working live is a last resort. Yet, I do not see how you can isolate the panels fully without a major shutdown of services. This will affect the building and have a negative impact on sensitive areas; furthermore, in places like hospitals, this really might not be an option.

Each contractor should have two first aiders and two fire marshals. You need to know where everyone is working. The permit system will only tell you the areas people work in undertaking activities considered risky.

For fire evacuation and other emergency procedures like floods and terrorist attack to work, you must have an attendance management system. This can be a sign-in/sign-out book, but it's a pain when fifty people come to your cabin daily to sign in and out . . . in fact, you could go mental after a week or two. One alternative is to obtain a fingerprint or card attendance management system. These can cost up to £1,000, but they will save you weeks of invaluable time.

Do not induct people without at least a basic DBS (criminal record) check and the compulsory asbestos awareness course, which is a legal requirement.

Request that managers provide you with a monthly H&S inspection report, so provide a form for that. The form should contain the records of weekly inspection of scaffolds, harnesses, MEWPs, ladders, stepladders and all other means of access. It is also a requirement to have a signed register of issued PPE. Free issue tens of near miss cards.

Run monthly H&S awards highlighting good H&S practices. Present gift vouchers to the best contractors and individuals. When planning for weekend cover, request that the subcontractors submit a list of personnel and supervisors who will be working on site every Friday by midday. Make sure you provide the cover yourself as a principal contractor. There may also need to have electrical and mechanical APs on site in case if they are required.

Risk Assessments and Method Statements (RAMS)

Personally review every risk assessment and method statement (RAMS). If you do not understand the works carried out, get a competent person to help you to review the works. I have yet to see a method statement and risk assessment where I did not make comments, asked questions or, worst of all, where RAMS are generic and copied from another job. Remember, if you allow works to commence, RAMS will be the first document referred to when accidents or a dangerous occurrence happens.

Your career is always on the line. Do not allow any pit bulls to pressurise you into allowing them to work unless you understand what they are doing 100%. Make sure that all the risks associated with the job are substantially reduced or eliminated. When it comes to H&S, you have all the power in the world, including stopping work, removing personnel off site and suspending work. H&S is not compulsory: it is law, and you as a principal contractor or contractor have a huge responsibilities and rights. No delay claims will be coming your way because of your strict or even harsh H&S requirements and demands. Do not feel guilty stopping progress because of poor H&S practices. It is better be strict than in jail in the case of serious H&S accidents or even death.

Reject generic RAMS immediately. Do not allow any works to commence until you are 100% sure that everyone understands what they are doing and have read, understood and signed the RAMS. Make sure that the people who actually do the job sign it. Make it a requirement for the RAMS to be written in conjunction with the operatives who will be doing the actual works. This is specifically important for works involving shutdowns, work at height, work in confined spaces or work on existing LTHW, DHW, steam and condensate systems.

You will often find it useful to visit the area where the works will be carried out. If you are still unclear, get the subcontractor to come on site and walk you through the works, risks and mitigating measures.

Finish the design, then build

Design and design development rarely seem to stop until the handover of the project. This costs you delays, lots of money and lots of aggravation. So, avoid the design stage dragging on into the construction phase, or worse, the commissioning stage. That's when you find out that stuff does not work.

Schedules of equipment for the chillers, air heat source pumps, AHUs, fans, control valves, meters, pumps, water treatment and conditioning equipment, pressurisation units, motorised damper actuators, plate heat exchangers should specify the exact location, technical data and the system served.

The designers can never foresee on paper the site's conditions, clashes with other services, structural problems with holes, complications and problems. That is why a proper site systems validation survey should be completed prior to any design work. This will eliminate the majority of problems, but it will never eliminate all of them.

Have a drawing register. All the drawings required to complete the project should be on the list. The first issue set will be preliminary. You will have to create a distribution list with all subcontractors, your client, maintenance and your team on it. Every drawing will need to be issued to your distribution list for comments. All the drawings, schedules, technical submissions, even contract documentation and specifications need to be reviewed, commented and approved by the people on the design review distribution list.

Have weekly design workshops. Keep them in good spirit. Create a design program that will tie up with the drawings register. A good practice is to invite experienced installation supervisors to these meetings. These guys will have a very different angle and input. Help them to come out of their plantroom comfort zone to the meeting room, as their comments might be eye opening.

Record and distribute the key points and comments from the workshop. The design engineers and consultants tend to run out of money and disappear after the completion of the design phase. New people will come on board during the installation and they will stick to the contracts and written agreements. If you do not record key agreements in the minutes of the workshop, you are unlikely to convince the contractors that the agreements were made.

You expect people to keep to their word. Bitter experience proves that unless you have a piece of paper proving the agreements, the response from the subcontractor will be a firm "No".

So record and distribute the key agreements from the design workshops to protect your team. Referring back to these at the construction stage will be invaluable in order to move swiftly forward.

If you are 100% sure that you have an unhelpful, unskilled, awkward subcontractor or other person, then remove him or her from the project without a second thought.

Good designers will want to finish the design.

Bad designers and consultants will drag it on for months. Never go to the installation phase until all the drawings are construction issue.

Appoint consultancies that take full design responsibility and have public insurance (PI). In the event that they make errors in the design, they will be liable for damages and rectifying the problem. You should determine this during the tender stage, as there are still companies on the market that do not take responsibility for their design: all they do is design development.

After the installation of the first plantroom or system is completed and snagged, set up another workshop with the designers and consultants. Review and accommodate the learning points, coordination issues, difficulties faced in terms of installation in relation to the drawings and specifications in the next set of drawings. There will be plenty of learning points. Issues will correlate to maintenance, operation, materials and accessories used, and access.

On the basis of these findings from this workshop, ask the designers to revise the drawings and design for the upcoming systems. No doubt you will have to instruct the designers, and it may cost you some money; however, it is a much better and cheaper approach than continuing regardless. Ultimately, it is better for the client, project and the installation team knowing that their feedback is being put to use and the systems will work much better.

The subcontractors will also come back to you asking for the variations based on the revised drawings. Remember: change often goes both ways. It may become cheaper on one side but more expensive elsewhere. Most importantly have a reserved budget allocated for the variations at about 25% of the project value.

BMS

Pay very special attention to coordination between the designers, mechanical contractor and BMS. The BMS project engineer must be involved in the design meetings from day one. The ultimate aim for the BMS is to make sure that every piece of kit requiring power is on the radar. This includes the meters, 24 V power supplies, de-aerators, pressurisation units, water softeners, stand-alone control panels and kit.

The ultimate document for the BMS design is quality description of operation. It is a document that describes how all the systems are controlled and monitored. The consultants and mechanical contractors, the client and FM team need to comment on this document.

In fact you should also have the maintenance team on the distribution list, as they will also have invaluable input into the access, maintainability and maintenance costs.

The best way of making sure that all the required power supplies are accommodated is by having a drive schedule. This will have a list of plant and accessories requiring power. The consultants will know what should be on the list. The mechanical contractor or principle contractor (if he purchases equipment) will provide the details on the consumption and connection details. The drive schedule should be created and driven by the BMS contractor, but it is your responsibility to make sure it is completed.

The BMS contractor must also create a schedule of control valves. The mechanical contractor will issue the flow rates, pressure drops and pipeline sizes for each of the systems. The BMS subcontractor will size, procure and free issue the control valves.

A schedule of fire and control motorised damper actuators is also essential.

The BMS contractor will also have to create a schedule of energy, flow, electric, gas, oil and water meters.

Mechanical

Make sure you see the schedules for the following equipment: pumps, fans, AHUs, plate heat exchangers, chillers, split systems/air heat source pumps, water softening/treatment equipment and meters.

For the mechanical wet systems make sure the drawings show the commissioning stations, drain cocks, flushing valves, temperature and pressure gauges, and BMS sensors with enough space before and after the commissioning stations to obtain accurate readings. Do not use double regulating valves as isolation points. Allow enough isolation valves so that the maintenance team can isolate parts of the system without draining the whole system.

If you are installing plate heat exchangers, allow the isolation valves on the flow and return with a flushing loop or valves for pressure testing and flushing. The manufacturers of PHEs do not want the plates to be pressure tested or chemically flushed.

It is a good practice to install some quality ball-type isolation valves on those key, large, critical systems: over time, butterfly valves leak water.

Screw-type connections are also guilty of causing leaks over time. The maintenance team are constantly patching screwed systems. Consider welded and flanged connections and accessories starting from 40mm. I know: mechanical contractors will make a fuss out of this requirement and it may cause longer delivery times for the fittings and accessories. Be firm and bear in mind that once the one-year guarantee expires, you will not get them back on site to fix leaks.

Do not allow MIG/MAG welding. All the welds should have a route. Set a welding procedure for steam, condensate, LTHW, CHW and connection methods for CWS and DHW. TIG welding is good for condensate and steam systems below 40mm. No screwed connections can be permitted on steam or condensate. Aim for stainless steel for the condensate pipework and SHED40 for the steam. Specify that the subcontractors provide the welding map for the project. This should include every single weld on the project. Specify that a qualified, independent welding inspector should inspect all the welds with a Non-Destructive Test (NDT).

Basis of a welding spec

The welds inspected by the certified and competent welding inspector needs to be agreed between the subcontractor and the principle contractor.

Items below are good practices and a good quality control system, which must be included in your contract with the mechanical contractors:

- ✓ Welding procedures to be approved by the principle contractor prior to the works;
- ✓ 100% of the welds to have NDT. This could be a mix of Magnetic Particle Inspection (MPI) and ultrasound inspection;
- ✓ 10% of all the welds to be X-rayed. It is unlikely that X-ray will be allowed on site. The welding inspector will watch each welder to do one-to-two welds. After that he will take the welded piece off site and then X-ray the welds. The results will be shared in the report. If faulty welds are found during the X-ray, another 10% of the welds made by that particular welder will be inspected. If more fail, 100% of the welds made by this welder will have to be inspected;
- ✓ A welding map to include every single new weld on the project. This map should clearly identify which welder made each weld.

Steam-system welding is a special customer and requires Class 1 welding. This means more stringent requirements for the welder's certification. It also means that in addition to the two bullet points above, the welds can be inspected against the welding procedure, which will also have to be approved by the principle contractor.

Certification

Water pressure testing should be around 6-7 bar for LTHW (or at least twice the working pressure as a minimum). The test should last 2 hours and extended to 4 hours if the pipework is already lagged. Air pressure testing is inconsistent and it's very hard to find leaks this way, so using it on site is not advised.

Certain parts of the systems, like headers or spools, could be chemically dipped rather than chemically flushed. Small sections of new DHW or DCW pipework can be also dipped for chlorination purposes.

Specify in the contract that it's the responsibility of the contractors to dispose of pre-commissioning flushing water from the site. This is because the contractors will be asking you to provide suitable drains. Make it their responsibility to dispose of the flushed water in accordance with environmental legislation.

Try to exclude provisions for the power supply and cold water supply for flushing as well. This will be used against you and you will be notified very promptly that you are holding the project up by not supplying power and CWS for flushing.

Insert strong coordination clauses between different contractors and designer into their contracts. This to include free-issue items like control valves, meters, pockets, connection and termination details for the BMS and power supply. The designers are also to produce fully coordinated drawings, showing all the mechanical and electrical systems, including the trays, panels, plinths, holes, builders works and location of the equipment.

Have specifications for LTHW, CHW, DHW, DCW, ventilation, lagging, steam and condensate, valves and accessories, pipework and ductwork and meters. You will also require individual specifications for the equipment starting from chillers, AHUs, PHEs, pumps, fans, electrical panels, BMS panels, cables, lighting and electrical installation. This is to protect yourself and the project.

Insert a clause about duty of care; with one of these, no one will be able to tell you later that this problem is not really his concern.

Pipework insulation

Effective pipework lagging will save energy, for H&S reasons, and also for protection against physical damage. As good practice, use the following:

- ✓ 100 mm of mineral wool for the steam/LTHW/condensate pipes over 80mm;
- ✓ 50-80mm of mineral wool for the steam/LTHW/condensate pipes below 80mm;
- ✓ For the areas outside exposed to rain, mineral wool to have both VentureClad and then 0.8mm AliClad protection on top;
- ✓ The brackets should have calcium or wooden blocks to insulate the brackets from the pipes—you do not want metal rods conducting heat;
- ✓ No silicon to be used outside to patch the holes in the AliClad;
- ✓ Step overs to be a part of the contract to protect the lagging;
- ✓ Consider netting as means of protecting the services against the birds such as pigeons, sea gulls and crows—their droppings are a serious H&S hazard and they do lots of damage to the insulation by ripping it off for their nests

When surveying the existing lagging of outdoors ductwork, pay attention to the condition of the ductwork lagging; this is very easily damaged by the elements. Furthermore, crows and sea gulls love ripping out PIB and VentureClad for their nests. Make sure the ducts are VentureClad and then AliClad-protected on top. Lagging sucks up water like a sponge and drastically reduces insulation efficiency.

Ductwork lagging, which hangs like a beer belly at the bottom of a duct, means it is full of water. Damaged lagging means only one thing: it needs full replacement. Patching up ductwork lagging is simply not going to work.

Design development 1: schematics and layout drawings

Installation and commissioning of the project will run much faster if there are quality, fully coordinated drawings produced. It all starts with the schematic drawing. A great schematic drawing should have the following:

- ✓ Drawings should be colour-coded to differentiate between what is being stripped out, what is staying and the new installation. When you read such a drawing, you should easily understand exactly what is staying and what will be replaced. The LTHW, CWS, DHW, gas, steam and condensate services should all have lines of a different colour. Colour coding hugely improves the readability of the drawing;
- ✓ If you do decide to produce a separate strip-out drawing, make sure you fully coordinate it with the layout installation drawing. There are plenty of examples where the new system has not been fully designed and the existing services had to be stripped out even further than the strip-out drawings show;
- ✓ All your drawings ought to have consistent line types and thicknesses, symbols, abbreviations, names, key or a legend system describing each of the symbols and elements on the drawing;
- ✓ The isolation valves on all of the services involved in a job have to be validated for operation. If the valves are not holding, the drawings should clearly state that those valves must be replaced for effective isolation. This is important from a H&S point of view;
- ✓ The layout drawing should be fully coordinated with the other packages, such as electrical services, BMS, ventilation, wet mechanical services, air conditioning, builders work (including holes and plinths), existing equipment and any other existing building services
- ✓ The drawings should show the containment for the electrical services and ductwork. Large ductwork should take priority during the installation, as it cannot be manoeuvred very easily

Under no circumstances should you start installation until you have fully coordinated construction-issue drawings, even if you are under huge pressure from your client to start on site or face program delays, as this will avoid any additional expenses incurred during installation.

I also strongly believe that starting based on preliminary-issue drawings takes much longer to complete the works. This is due to coordination and installation issues that will almost certainly surface. It is impossible to resolve them fast and effectively when you have twenty installers waiting to crack on with installation or where you need to arrange yet another heating shutdown in the middle of winter.

Design development 2: technical submittals, RFIs, description of operation and installation drawings

Appointed installation contractors will be submitting technical submittals. The submittals will be based on the specifications and their preferences in terms of the product, installation method and the cost (normally the cheaper ones are proposed).

To clarify, the specifications will not tell you to buy Schneider Electric's switchgear or Crane valves; it is the duty of the installation contractor to select and propose suppliers. If the specification is vague, or there is no section in the specification for this specific material, the contractors will make their proposals.

The contractors will often want to use cheaper alternatives and faster installation methods. To make sure you get want you want, the technical submittals should be reviewed by a review panel (the people on your distribution list). As with any other design documentation, the review panel will consist of the designers, other contractors involved into the project, the client's project manager, the principal contractor and experts from the maintenance team.

Those on the review panel will be reviewing the submittals. Every single piece of equipment, its fitting and method of installation will be selected and a data sheet submitted for approval.

Before splashing your cash on anything, make sure you have a technical submittal approved to at least Status B (approved with minor comments). Otherwise, you expose the other contractors, client, designers, principal contractor and maintenance team who will inherit the systems after the installation to huge financial, commercial, reputational and operational risk.

The technical submittals are a critical part of the design. Study and review them carefully. Watch out for coordination issues, optional items, extras, interface cards, and make sure you have all the information required. Submitting a data sheet or catalogue without specifying the model and any associated extras is not good enough. Each technical submittal should have a cover page describing in detail the model and all optional items included. Often, the contractors will drop all the extra options, as they cost money.

Wiring diagrams with power and control sections should be a part of the technical submission.

BMS design

Bear in mind that your electrical or BMS contractor will be manufacturing the panels, selecting breakers and switchgear based on the information in the technical submittals and producing the wiring diagrams with the termination details. It is absolutely important to get it right. The consultants and BMS designer have to pick up items such as extra interface cards required for the equipment to be controlled effectively.

The equipment should have the capabilities to be controlled by the BMS as per description of operation. The Description of Operation is the most important document of the BMS design, so do not take it lightly. For example, if you have built your panel with 0-10V output, but the boiler can only take 4-20mA, then you have a problem that will cost you a panel modification. However, if you know it in advance, then only the wiring diagram needs to be changed.

Another common thing that happens is when the mechanical contractor buys three stand-alone chillers and misses that the spec is calling for the chillers to come with sequence control panels. You can't expect the BMS to sequence the chillers at no extra cost. Site instructions come out, the design is delayed, and the BMS panel is modified on site. Ultimately, it takes longer to install and commission the chillers. The fact is that everyone involved in the project loses through laziness, incompetence and lack of cooperation during the design phase.

When reviewing the technical submissions, pay attention to details.

Similar mistakes to the above chillers are usually made when buying, AHUs, boilers, skids, plate heat exchangers and even pressurisation units.

The BMS drive schedule is another must-have document during the BMS design. This usually ties up all the loose ends on the power side; it will include every item in the area requiring power supply, details of the electrical load and connection details.

Items like water softeners, de-aerators, meters and all sorts of stand-alone skids are often missed out. Based on that drive schedule, the BMS contractor will be able to select the panel gear and size the cables.

Request for Information (RFI)

Every single piece of information you generally require should be submitted formally as a Request for Information (RFI). The best way of managing the RFIs is to create a schedule of RFIs. It is a great tool to suss out which piece of missing information is delaying the project's progress. Make sure you do not abuse the RFI system and talk to your client. Do not ask about something that is already in your contract documentation.

Installation drawings

Based on the construction-issue layout drawings, the mechanical contractor will be producing installation drawings. These drawings will show the spool pieces for the pipework so the welders can prefabricate them. The same applies to the ductwork, which will be prefabricated off site. When it is shipped to site, the fitters will have installation drawings showing them how to assemble the sections in sequence like you do with Lego. Based on the installation drawings and technical submittals, the contractors can order the materials, brackets, gaskets and pre-fabricated pieces for the ducts and pipes.

Selecting and buying equipment with a long delivery time

Your program will be heavily dependent on the delivery of key equipment to site. AHUs, plate heat exchangers, chillers, pumps, pressurisation units, ball valves, steam valves, boilers, low loss transformers, steam traps and any off-normal specification items might have a delivery time of up to two months. CHP and biomass boilers may take up to 6 months to manufacture. The point is that you cannot even buy them until the equipment is selected correctly. The equipment we are talking about here is very expensive, so you definitely do not want to select the wrong ones.

You have to go through the rigorous process of the equipment and accessories selection and review, which takes time that you will need to account for. Your client and people on your design review distribution list will have to see and comment on the technical submissions and drawings issued. The comments might take up to a week or two to produce. Then, again, you will have to review the comments, which can then have effect on the price. The quote from the suppliers will have to be revised. Then, you will have to push hard to raise the purchase order though your commercial manager or quantity surveyor to get it through the system.

For big orders, nowadays you will need MD or SEO approval. Again, it can take weeks, if not months, to get that sorted. The problem might be that at the time when your program dictates you placing the order for the long delivery items, you might not even have a purchase order from your client; yet, your client expects you to honour the program and get on with the project regardless.

I have yet to be on a project where the works weren't held up by the long delivery time of some key equipment. Until the equipment is on site, there are always delays with the installation.

Because on-time equipment delivery is such an important part of a successful project, let's look into this in more detail. First of all, make sure that you have an up-to-date and accurate quote from the supplier for this equipment. Very often the supplies are still awaiting some clarifications on the design parameters. Make sure there is no hold up here. Get the required, and accurate, information to your supplier's technical department as soon as possible.

Once the quote is received, make sure you fully understand the terms and conditions. More importantly for the project, make sure you place a full order with the supplier and that the sales rep from the supplier has a copy also. Your bargaining period is just before sending the order. The time just before placing the order is an opportunity for a special discount if you place the whole order at once. More importantly, try to agree an accelerated delivery time for the equipment, which is even more important for you.

A good practice is to have Factory Acceptance Tests (FATs) just before the equipment is shipped to site. Invite the specialists from other specialities, your client, maintenance manager and your design consultants to the FATs. BMS is one area that specifically springs to mind. Make sure, that the kit is up to spec and can be controlled by the BMS, and especially if some coordination is required between the communication protocols.

Aim to have proper functionality tests during the Factory Acceptance Tests (FATs) during which time you should pay special attention to the special systems.

Screw-ups may also happen with the deliveries to site where lorries turn up without hi-ups, panels without lifting eyes, or when a crane is required to lift the AHUs, chillers or the boilers to the roof but the crane has not been arranged with the client.

Also, do not forget to sort out and complete all the required builders works in conjunction. If you need the plinths or holes, make sure it is all done beforehand.

In terms of the crane, always check the lifting plan. Make sure that the structural engineer surveyed the drop-off area and the area where the crane will be rigged. This should involve sample drilling based on the site infrastructure plan, which you will have to obtain. Sample drilling might have to be done outside of working hours. Bear in mind that all the lifting companies will exclude their liability for damages to the infrastructure under the crane or in a drop-off zone.

Shine and glow. Tips for project managers

To become a great project manager, you should invest years into developing the following essential key skills:

1. Interpersonal skills

This is an ability to understand and drive the types of personalities of the project's stakeholders. It is about the skill of adjusting what you say and how you act to opposite personality types. At its most basic, a team can be broken down in to 4 personality types and denoted by a particular colour: red, blue, green and yellow (reference taken from http://www.quia.com/files/quia/users/kkacher/WrldHlthResrch_handouts/Personality-Test-for-Teaming).

The reds are the drivers and completers. For them it is all about completing the task or making sound progress. When they are under pressure, reds can often become even more "red" by being pushy, direct and may even seem untactful and authoritative.

Yellows are "people's people". They get on with others and are loved by everybody; we usually say he is a nice guy. They get on with everyone. For yellows, people need to be happy. That is their world. There is no other way and there is nothing more important than being happy. Yellows are the glue of the team and are consummate team players. They are an asset to any team and project. They create a happy environment to work in and work collaboratively. To convince the "yellow" to do something, highlight the team benefits.

Greens are the techies (nerds). They like spending time on their own with gadgets and Excel spreadsheets. They are great with computers and have a brilliant technical knowledge of the systems and technologies. To convince them, give them facts, figures, charts, statistics and data. And another thing, never ask a green person how he feels about the project: greens don't do feelings. Questions like these will set them on a never-ending algorithm of trying to work out what "feeling" means. They are not good at feelings or showing them. As with any of the above personalities, the greens are also essential for any balanced team. They will pick up faults with the specs, drawings and will snag the job to death.

Blues are creative and innovative but reserved and do not run with the pack.

Myself, I am situated between red and yellow; however, under pressure and stress, I shoot from mid-yellow and red straight into the red. So, I understand how frustrated the reds get when things drift off course and no progress is being made. We take that very personally and start impatiently fighting for the cause. Do not confront reds head to head. It is pointless. It is the way they see the world. It is their reality and for them this is the only right way of doing things. They either get things done or burn themselves out. All great project managers recognise when they approach a red's area and back off or start acting smarter.

The perfect team is a balanced team with all the above personalities involved. Of course, it brings its problems at an interpersonal level, so every person on the team needs to adjust their communication style knowing the strengths of the personalities and personality types.

A great project manager intuitively builds a balanced team of all personalities and influences each personality.

2. Unstoppable drive for the project's progression

No matter how great you are at the rest of the important skills needed for successful project management, if you do not have the drive for the project's progression, the project will fail or be mediocre at best.

You should develop a constant and burning desire to progress, drive and direct the project. Making no progress should hurt like hell. You should be constantly getting to the bottom of issues, working out (with help) how to overcome these problems, picking up the phone, meeting people and resolving the showstoppers.

*Great PMs are very comfortable making uncomfortable decisions and
completing the items we dread the most.*

No issues should be holding up your subcontractors. You need to develop the skill of cutting through the excuses straight to the heart of the problem whether it is labour, supervisor, holdup of materials, unfinished design, etc. Great PMs drive the job through.

3. Competency is building services engineering, installation, commissioning and maintenance

Normally, project managers grow out of installation supervisors, engineers or commissioning managers. The project manager needs to understand building services, how to design the systems, how to install them, how to commission them and how to maintain them.

No one should be able to pull the wool over your eyes. You need to be able to make your own opinion about the situation very quickly using your practical knowledge of the design, installation methods, and quality of work and commissioning. When you examine the drawing, specification, schedule or a technical submission, you should be able to judge very quickly whether the drawing is of an acceptable or unacceptable quality and to be able to provide dozens of comments.

When you look at the procurement schedule, you need to be able to understand, realistically, how soon the installation will commence on site. When you walk around the site, you should be able to pick up snags as you go along and be able to point out the practices that do not comply with the specification or the construction drawings; the same applies to the commissioning, testing and balancing. You need to know it all, because if you do not, then the project is at risk.

In simple terms, you need to know your stuff. Now, of course, all of us have some disciplines in which we are strong; however, it is essential to have a very good understanding regarding every step of the project or to have team members in your project team covering all the aspects of the project from specifications to the commissioning witnessing.

4. Project authority and having a high profile

It is not my personal favourite, but a great project manager projects authority in any situation. Britain loves authority. It is imbedded in the psyche. Maybe this comes from its imperial background with its aristocracy. If you do not have authority, you will find it very difficult to manage the project and the customer. This is because every step and decision you make will be judged, scrutinised and criticised. Unless you project authority, even if it's acting, you will not pull it off.

You need to be comfortable chairing and controlling a meeting with the subcontractors and the customer. You need to have the confidence to be fair and firm and at the same time flexible and accommodating. Successful project management is a balancing act between the project's progress and customer satisfaction. You need to be successful at both.

Authority generally comes with age. To gain authority, you normally have to be over 40, have grey hair or at least a balding head (these simply make you look older). You should also have an important title, like Senior Project Manager, Director of Projects or similar. And another thing, you will have to be dressed for business: a corporate polo shirt won't cut it.

So build you career in your organisation by being mates with the right people, being politically aware, and by delivering stunning projects. The grey hair, balding head and the job title will follow.

5. Communication skills

Communication is another important skill, but it's often misunderstood and confused with interpersonal skills and a supreme knowledge of the English language. Communication is all about making sure that the message you give is understood.

So often we are conditioned and attuned to hear what we think is important and wiping the rest of the message. The best way of making sure you have understood someone is to repeat the message in your own words. If the messenger agrees with what you said, then the message was communicated successfully.

6. Influencing skills

Influencing is a very important skill. Partially, it is about the dirty tricks used in TV adverts, some of which use the foot-in-the-door technique, the principle of consistency, the principle of scarcity, and the principle of association. You need to be able to direct your troops and more importantly drive and manage your customer.

Your actions, timing, language, tone and your approach influences the client, suppliers, subcontractors and your project team to act in the direction of successful progression.

Summarising, fantastic project managers are not born, they are made. From project to project, they continually develop themselves by modelling the great skills of other people. They continue going even after making painful mistakes and stay on course. Aim at developing the skills mentioned above throughout your career. Work for and with the people from whom you can learn. Stay on the path of mastery and enjoy the journey!

You should be on the job either making great money or constantly learning and improving. If you feel that you do neither of those, should you not shake things up in your career?

PART 3

Construction

Installation program

If I may, I would like to coach you on how to create an accurate installation program. But first, let's get the hang of the amazing and satisfying feeling of being in control of a successful project.

Imagine you attend the weekly progress meetings with your client. You sit in the meeting room very comfortably. You are confident and happy. Your customers thank you for achieving the milestones on the program. They thank you for delivering the elements of work ahead of schedule. The customers ask you how you planned everything so precisely, and you answer with pleasure and poise.

The best time to pursue the installation program is towards the end of the design stage. The purpose of the installation program is to focus on planning and organising the construction phase of the project. In that sense, it is different from the overall project program.

The only right way of doing the installation program is starting from the end.

Step 1.

Get the sub-subcontractors to do their own installation programs, focusing only on the duration of the tasks (no specific dates).

Do not create the program for the subcontractors: first of all, they will not commit to it; secondly, you will overestimate the speed of works. In the end, you can shout at the subcontractors when things slip up; you can demand that they squeeze the duration or even accelerate works; you can also threaten them with liquidated damages.

However, at the end of the day, it will take them three times longer to complete your works than you have programmed and anticipated. You might not be happy about it, but that will be a true duration. So, stop deluding yourself. It is time to get real.

I would like to stress another crucial point: as you are asking your subcontractors to produce the program of works, your subcontractors should be asking their subcontractors (your sub-subcontractors) to produce their installation program as well. The thing is that you need to understand how your subcontractor is planning to deliver the project for you. If the subcontractor offloads the works to his subcontractors, then what is the value in having such a subcontractor? You could employ a small management team to manage that package and hire the sub-subcontractor directly.

Understand this: your subcontractor is, in effect, an agent of their subcontractors or your sub-subcontractors. They do not call the shots. Despite all their bravado, ambition, status and experience, all your subcontractor will be doing is passing the information from their subcontractors to you and back again. The problems of your sub-subcontractors will become your headache. The bad thing is that when the package is not performing, you will have to deal with and manage your sub-subcontractors. I often ask myself why am I paying huge preliminaries and margins to subcontractors if they aren't managing their subcontractors?

Once you have all the programs from each of the subcontractors, double check that these programs were based on the programs of the sub-subcontractors.

Step 2.

To verify the accuracy of the program, request procurement schedules with the expected delivery dates and confirmed by the suppliers. The procurement schedules should tie up with the program.

Also, check the program against the schedule of specifications, schedules of equipment, drawings and technical submissions. Installation on site cannot commence until the specifications, drawings and technical submissions are approved. The subcontractors often forget to check these and overcommit.

If there are any issues with resourcing the job, then you need to fight for the best resources the subcontractor's organisation can provide. You can only do this by fully understanding the project and by escalating issues straight to the decision makers (Vice Presidents, Managing Directors and CEOs).

I do not like to use the misused phrase "apply pressure" to get things sorted or the even worse one called "put under pressure": I do not believe in these principles. Let me explain. Paraphrasing Pascal's law from physics, pressure creates equal resistance. Excessive pressure, aside from creating excessive resistance, can also alienate and demoralise.

However, I like the phrase "apply necessary focus", because it is a smarter way of achieving the desired outcome. In order to achieve the progress you will have to know what the bottom-line problem is.

As your contract is not with your sub-subcontractors the subcontractors are prone to go a long way in hiding what the real issues are, trying to eliminate any direct contact with the sub-subcontractors and use all the bad tools possible in finding why they can't do the job at the rate you want.

It is the game the subcontractors are used to. Under pressure, they have mastered a way of finding what is holding them up. The attitude goes like this: *"The answer is no. What was the question? It is not ours. No one tells us anything. We can't do that. The notice is too short. It is not in the contract. You, as a principal contractor, need to do this. It is not our responsibility. They have not done that! You see, no one tells us anything here."*

What that waffle generally means is that the subcontractor does not know yet how to achieve what you are asking. He is buying some time to find an excuse and seeing if he can get away with it.

The last pieces of the installation program puzzle are:

> ➢ the duration of mobilisation of the required labour, management, supervision;
> ➢ duration of the site set up, including the cabins, store, toolboxes, means of access;
> ➢ having all of the H&S aspects of the job sorted, including the Risk Assessments and Method Statements (RAMS) and safe systems of work.

Each of the above mentioned items are critical for the project success and can't be overlooked.

Step 3.

Draft a master installation program that includes all the subcontractors' works.

This is something you need to do yourself. Examine the sequencing of works. Make sure that works go in a logical way without clashes and held ups, i.e. builders works in conjunction, mechanical and electrical, BMS and pipework insulation.

Step 4.

Arrange an installation program coordination workshop with all the subcontractors involved and revise your master program based on the outcome of the workshop.

The purpose of this workshop is to coordinate the programs: for example, to make sure that there is power for the pumps for flushing, cranes are arranged, structural surveys are booked, that the builders works are completed before M&E works can commence and that the working areas are not overcrowded with people.

During the coordination workshop, it is time to look at the items sitting between the contracts of your subcontractors, as that is where you are likely to trip. Ask your subcontractors what they foresee as showstoppers, problems and potential issues. Allocate time in your master program covering all the packages and issues that surfaced during the workshop.

Step 5.

The next step is to book another workshop. This one is with your client and designers.

This is to go through your coordinated installation master program:

> - check for clashes with other capital projects;
> - coordinate an adequate period of notice for systems shutdowns with the maintenance team;
> - coordinate timing with third party works.

Step 6.

Only after the coordination workshops, firstly with the subcontractors and then with your client, can you can make the necessary adjustments in your program and issue the master installation program to the client, subcontractors and project team.

Arranging systems shutdowns (micro schedules)

Believe it or not, it's a lot of hassle to arrange a shutdown of any services. To name a few, the heating system, DHW, DCW, AHU, lighting circuits, small power, distribution boards, steam system, BMS, gas, lifts and compressed air.

Your client is lazy, stressed, overworked and overloaded. He will not want to make the effort of running around and finding out what those systems feed and what the full effect will be of a full systems shutdown. He, however, is the person from the capital projects team who will be authorising the shutdown. With responsibility, however, comes blame, and especially when shutdowns cause complaints or works do not go to plan.

You can offload this hassle to your client, but do not hope for a quick "*Yes, you can do it*" answer. To understand why, you need to comprehend how estates and facilities departments work. For large estates, the FM team will be split between maintenance and projects. Your client is a project manager from the capital projects team. Maintenance, through the helpdesk, will be acting on any complaints that your shutdown causes. They will go out of their skin to make sure that projects team does not cause their already understaffed, not fully competent and stressed team additional problems. In other words you are a target here.

Every wrong step you make, every shutdown you make which causes even the most minor of inconveniences will be escalated and the project team will be beaten up for it.

The estates and facilities' function is a supportive one: they are there to support the main function of the business, i.e. to teach, heal, deliver a product or service or make money. Every single thing that causes even a minor disruption to the main function of the business will be criticised and condemned.

Now that you understand the full implication of any of the shutdowns required for you to complete your contractual obligation in delivering your project, you will be spending more time planning the works and liaising with people and the end users.

Attend weekly maintenance meetings to put forward your proposals for shutdowns.

Also, do not forget that you have a program to honour. If the shutdowns do not happen, your manager will remind you that you do not have forever to deliver the project.

Summarising, you need to take on the responsibility for planning and executing the shutdowns into your own hands to get it done.

First of all, understand which areas are fed from the distribution board, heating plantroom and BMS panel. It might sound silly, but in practice the electrical panel outgoing feeds are likely to have old and meaningless references, if any, with no references to where the power is coming from. How are you supposed to conduct an earth-loop impedance test and complete the test sheets?

The heating pumps may not have any labels at all. The systems may have been extended with no records of why or where. You might find yourself in the peculiar situation of tracing the pipework, ductwork and cabling and speaking to the most knowledgeable maintenance guys on site.

The funny thing is that knowledgeable personnel are either constantly being made redundant (too expensive), leave their current jobs for better money or retire. There is a huge services knowledge deficit in the current FM business. New guys come in with the wrong attitude and with no desire to learn the ropes. Cherish great FM guys. Go the extra mile in building relationships with them, as they have all the knowledge required.

To help you to comply with industry guidance, shutting down the systems is regularly required for inspection, testing, maintenance, checks and insurance purposes. You might need to get through a few barriers of tough resistance before you get the information you need. I bet you did not think it was your job to find what is fed from the existing plantroom. However, if you want the project to succeed, you will have to deal with that issue as well, so roll up your sleeves and act. It is not a new-build where you can simply ask the designers for guidance.

Once you know what the systems are feeding, understand the full impact of the proposed shutdowns. Again, you have to roll up your sleeves, find out the contact details of the departmental heads, lab managers and whoever is in charge of that area. To get things moving quickly, you are likely to talk to those people and find out when is the best time to switch off their heating, DHW, ventilation or power. It will very likely be outside of normal working hours, so be prepared for that, because some of your subcontractors will ask for double time and will want an instruction to that effect.

You will be lucky if your client can deal with all of those issues for you and simply tell you the date and time when to plan the shutdown.

System shutdown micro-schedule (mini-program)

Once you know when you can shut down the system, the next thing that you will need is to create a micro-schedule (mini-program in Excel) of your sequence of works showing the duration and timing of each task. As an example, let's say, you will:

- ✓ Isolate LTHW at 08.00 on the 12th of September 2015 and let it cool down for about 2 hours until the temperature is below 40 °C;
- ✓ Drain the system;
- ✓ Disconnect the electrical supply to the pump;
- ✓ Take the old pump out, put the new one in;
- ✓ Re-connect the electrical cable, check the rotation;
- ✓ Complete electrical test;
- ✓ Fill the system and run it to make sure that there are no leaks and that the heating is being circulated;
- ✓ Vent the system and radiators if required;
- ✓ Check the BMS periodically for comfort parameters and regularly report the progress of works via email;
- ✓ Work scheduled to be completed at 15:00.

In our example, you might also need to:

✓ Install some temperature loggers in the room space affected by your works. Do this before the shutdown and then monitor the temperatures during the operation and for a few days after;
✓ If the space is occupied 24/7, you might need to add some electrical oil filled heaters to make sure that the temperature does not drop, say, below 20 °C;
✓ Monitor the outside temperature as well to profile and predict its effect on the space temperature for future shutdowns;
✓ Have someone visit the areas every three hours to talk to and update the building's staff and visitors.

If you are planning longer shutdowns, say for few days, then make sure you survey the electrical infrastructure beforehand to ensure that it will take the additional electrical heating load without knocking out the power for a whole floor.

Following up on our example, make sure that each of the tasks described has a start and finish time. Creating this schedule proves to your client that you fully understand the implications of the works you are doing and that you have taken all the necessary measures to eliminate or reduce the risk of further disruption. No one wants to take a financial and reputational hit caused by an unplanned disruption to his or her business. You certainly do not want to be the subject of newspaper stories about causing yet another service disruption in public services.

Issue that micro-schedule (mini-program) to your client for formal approval and authorisation of the shutdown. Run through it face to face with the departments affected, your client, the maintenance team, and accommodate any comments and feedback.

If required, arrange maintenance tickets for isolations and obtain the permits to work from your client.

Electrical and mechanical isolations

You will have to put in place, and then run, systems for electrical and mechanical isolations.

Starting with electrical isolation, when your mechanical contractor wants to strip a piece of decommissioned electrical equipment, you will have to make sure that it is electrically dead and isolated (disconnected) from the power source. The correct way of going about it is to have an almost full-time appointed person (AP) on site who is competent with electrical isolations and energy equipment.

The AP will run an electrical isolation permit system. He will make sure that the equipment is dead and disconnected. He will also padlock the feeding circuit breaker so no one can accidentally energise the circuit while the mechanical contractor is stripping out the equipment.

A similar process will happen when the mechanical contractor needs power for flushing, power to the pressurisation unit or a pump for commissioning. The AP will put in place a request-for-power form. To turn the equipment's power on, the mechanical contractor will have to complete one of these request-for-power forms and specify the equipment requiring energy. The mechanical contractor might be required to give, say, 48-hour notice so that it can be arranged.

In this case, the AP will make sure that the piece of mechanical equipment is correctly installed, the wet system is full of water (this is to avoid burning pumps shifting air instead of the water, as the pumps are unable to cool down without water passing through them), and may be even pressure tested. Once the AP is happy with the above, he will energise the equipment and may even put a notice in the plant room that the equipment is now live. Being an AP can be a full-time job, as it involves lots of work and paperwork.

Mechanical isolations

You will also need an Appointed Person (AP) for mechanical isolations (a different person). Sections of LTHW might need to be drained, and from the health and safety point of view, you must ensure that people will not start cracking the joint of a 2.5 bar system with 80 °C LTHW and consequently scalding themselves. As with the electrical isolation, the AP will have to run the mechanical isolation permit to work. Your mechanical subcontractor will not be able to isolate anything until your AP has signed the permit-to-work for that task. A similar process applies to steam systems; however the AP will need to be formally trained to work on these.

The AP will make sure that the isolation valves are isolated and holding; he will then chain and padlock the isolation valves so that no one apart from him can open them.

If the isolation valves are not holding (water continues to escape) the AP might need to go further up the circuit to find another point of isolation. In a worst-case scenario, the whole system might need to be stopped and drained.

Bear in mind that although the system could be isolated, it will still have static pressure, as it is a pressurised system. So, be very careful when cracking the joints or undoing flange bolts. The AP needs to be present during this procedure to see that the work is done safely. The water in the system needs to be drained somewhere, so make sure that there is a functional gully nearby. Use a water hose to get the water to a gully and never assume that the plantroom floor is waterproof. Cases of flooding the areas beneath the plantrooms might sound unbelievable, but they happen more often than you can think.

Oftentimes the old system won't have enough functional valves and you will have to drain the whole system. Make sure that the pressurisation unit is off; otherwise, it will kick-in to fill the system with water. All pumps must be switched off and the system allowed to cool below 40 °C.

You can now see how the simple task of changing a single pump can become a huge and painful undertaking.

An alternative way of draining certain sections is pipe freezing; but again, the pumps will need to be shut and the water allowed to cool. In this case, you will have to produce a micro schedule for the shutdown so the customer can approve it with the building's users.

For steam systems, your customer will isolate and chain the steam and condensate valves first, and then your AP will padlock the system again when he is satisfied with the isolation. After this, the subcontractor who is actually doing the works might put on a third padlock. This is best practice and ensures that the unfinished system cannot be turned on without the consent of all three parties.

The steam system needs double isolation, meaning that to shut down the calorifier, two steam valves need to be isolated and chained. The AP has to make sure that both valves are holding 100%. To check that the fist isolation valve is holding, the AP normally opens a small bleed valve between the isolation valves. If no steam escapes the open bleed valve, then the AP will be satisfied with the double isolation. Condensate does not need double isolation.

With steam systems, the permit system should be cascaded down from the client to your AP and then from your AP to the subcontractor involved.

Area acceptance document

You might be wondering what an area acceptance is and when it is used. To clarify, prior to starting any work in the working area, such as the plant room, roof or a service duct, you need to spend some time in the area making sure the area is ready for your works to start. There is no point in you committing to complete the works if the plant room is used as storage. Make sure that the storage materials are removed beforehand. It should also be clean from rubbish and debris. If you start your contract work in this area, the progress will be like walking in a minefield. As soon as you start, all that rubbish will be your responsibility along with a request to keep your areas clean and tidy.

There might be simply not enough space for your lads in the room if there are other contractors working in the area. Record this in the area acceptance sheet (see next page) and issue it to your client. You are now officially being delayed. If you are planning to use drains for draining pipe work and flushing, make sure they function. There are two ways you can do test the drains: ask via an RFI and speak to the maintenance team, or you can test it yourself and see what happens.

If you submitted an RFI (with the picture attached), you are then covered and not liable for any flooding of the floors below. State in the RFI that if you do not receive an answer in seven working days, then you will assume that the drain is fully functional.

The same applies if there are existing maintenance problems in the area, such as leaking pipes and valves. These should be reported in the area acceptance sheet. If the customer doesn't have the time or money to repair the minor leaks, then at least this is recorded and you are less likely to be blamed for causing those leaks. And believe me, you'll be blamed for any and all maintenance issues associated with the areas in which you work. You will become a part-time maintenance helpdesk fixing everybody else's problems and sorting out what caused the problem. This is, of course, unless you invest your time in completing area acceptance documents.

Take plenty of pictures of the current issues, missing lagging, damaged equipment and damaged lagging and attach all of them to your report. This is your evidence and the only way of protecting yourself. It sounds negative, and not in a spirit of the team cooperation, but it also protects your client who is likely to be a project manager from the capital projects against the claims from the maintenance department.

For the same reason, if there are certain aspects of the contract that are down to your client to complete, like providing the power supply for welding machines, builders work or any other enabling works, do not hesitate to add them all to this area acceptance document.

If you think your customer might take these issues personally, the best way is taking your client around and visiting these areas and pointing out these issues. Once the issues are agreed, only then should you issue the document.

Use the area acceptance document below to help you out.

Item	Description	Date	Status, y/n
1	Area is clean from rubbish, debris and stored materials (pictures taken)		
2	Area is not overcrowded with other contractors to the point where it is impossible to work without a H&S risk		
3	Any existing maintenance issues, such as water leaks, leaking valves, missing actuators, ceased pumps, faulty lights and non-operational equipment and accessories, are resolved and repaired. If not, take plenty of pictures for evidence		
4	Safe and adequate access and egress is available		
5	Any damaged lagging, cladding, equipment, fabric, floors, walls, ceiling, doors or windows is recorded and take plenty of pictures as evidence		
6	Confirm via a Request For Information (RFI) that the plantroom floors are sealed and the gullies and drains are fully operational		
7	Record any operational and maintenance issues on the systems you are about to refurbish. If you don't, historic issue will become your problems. This includes non-functional isolation valves, leaking roof, no identification or labels on equipment		
8	Record any unsafe and high risk conditions and get maintenance to resolve them prior to start (dead rats, cockroaches, pigeon droppings, no pigeon netting, other contractors working with no PPE, no edge protection)		
9	Enabling works (if not part of your contract) are completed:		
	1) holes in the ceiling, floor or walls;		
	2) ceiling tiles removed;		
	3) temporary power supply provided (welding machines, power for flushing, etc.)		
	4) plinths are installed		
10	Any other related observations		

Supervision, labour, then management

Getting a top-notch supervisor should be the first thing, followed by quality labour and only then a manager. This idea comes from experience.

A great manager with an unskilled supervisor and agency labour equals a disastrous project.

Great labour *without* a top-notch supervisor equals inefficient and frustrated labour constantly waiting for materials, equipment, means of access and not being able to perform their task effectively. This also means top resources leaving your project for one that is better organised; a project where they can do their job and take pride in their work.

Get outstanding supervisors. An outstanding supervisor is the most important ingredient for a successful project.

Labour and supervision

A fantastic supervisor comes "from the tools", is great with people, has a head for design, understands installation and maintenance, and is great at keeping the site safe. An outstanding supervisor will have successfully delivered similar projects in the past. So start the recruitment of your top project team by appointing a great supervisor, preferably someone you know personally and get along with very well.

The supervisor is the person who will be clearing the path for the workforce so that your installers, welders, pipe fitters, electricians and specialists can do their job easily and with grace. It is the role of the supervisor to make sure that the site is set up right for the project, that the design is completed prior to the installation, and that all the tools and means of access are hired and ready for the guys to crack on with the job. It is his job to liaise with the client and deal with the numerous small issues that inevitably arise on site.

A top supervisor will be planning for the future to make sure that all the materials are bought to spec and on site on time. He will maintain a great H&S site record by sorting any issues straight away.

Here's the important message: building services contractors no longer employ quality labour directly. There are no companies that have great welders, pipe fitters, BMS commissioning and electricians working for them on the books. All the best skilled labour is free to choose which job to take on and how much they are to be paid. How will you make sure that you get professionals for your project that are good at what they do?

Great supervisors who came from the tools and have been around for years know the best guys and certainly do not want cowboys working for them. They will help you to select the right resource.

You should have a supervisor for each discipline: electrical works, mechanical wet systems, mechanical air systems, lagging and BMS.

If steam works are involved on site, then this will be totally different from normal LTHW works, as it requires Class 1 welding and the majority of welders will not be qualified for these works. Steam is unique in many ways and therefore requires a supervisor who has worked with steam systems before.

Great supervisors will be spending almost all of their time on site with the guys. You will see them explaining to the workforce what to do and solving their small problems: for example, where existing isolation valves do not hold. He will be in charge of logistics and have a role as part-time store man, unless the job is so big that it requires a full-time store man.

The supervisor will make your job more enjoyable, as you will not even know about the hundreds of issues he resolved today and will resolve tomorrow. The job will be running nice and smooth. You as a manager will be operating at peak level when the supervisor does his job. You will not need to get bogged down in micro-issues. You will have time to manage the project. You will have to go through a number of jobs before spotting star supervisors, but they are as good as gold. Once you find them, look after them and pay them well to keep them.

Management

Get successful managers to work for you.

Each reasonable package will have a project manager and a site manager. Let's talk about making sure you recruit the right ones as well. It is very unlikely that your subcontractor will appoint a project manager and site manager whom you already know. Also, unfortunately, it is impossible to judge the quality of the management based on the first few encounters until you are some way into the project.

First impressions often do not count, and you will often need to give a person up to two months to show how good he or she is.

Judge the project manager by excellent progress and results.

Never judge a person based on their talk. The more skilful a project manager is at talking and sounding like an expert, the worse it is for the project. Do not put up with the consistent underperformance. Show your strength and remove underperformers from the project. If you do not remove underperformers from the project, you will regret it a thousand times later, as all your energy and focus will be sucked into micro-managing the package that he or she should be in charge of.

You might have to escalate the matter some way up in the subcontractor's organisation and be very firm and persistent in replacing the project or site manager. You can only do so based on their performance and often involves commercial leverage and exercising your will and determination. More often than not, you will have to prove the point that you are the boss.

The less happy you become with the progress of works, the more influencing, wise, sometimes authoritarian and demanding you will have to become to get the project moving again.

Site set up, mess facility and storage

This seemingly unimportant chapter becomes very important indeed if you get it all wrong.

Here is the best practice.

For starters, the subcontractors' offices must be in the same location as yours (maximum 20 meters away from you). This will dramatically reduce the amount of email communication (awfully bad for your project), build a stronger relationship and ultimately get better results. Use all your influence and wits with your client and get a spacious spot on site to set up cabins. Splash out and supply cabins for your main subcontractors and position the cabins a few meters away from each other. Have separate cabins to give people private space. This is so you can still have a private conversation with your team without the subcontractors overhearing.

Under CDM Regulations, as principal contractor you have to provide a mess facility. Something basic like a large Portakabin, or other portable cabin, with tables, chairs, microwave, hot and cold water, fridge, kitchen towel, some storage cabinets and a kettle will be a perfectly good place for site personnel to hang out during breaks. Provide plenty of rubbish bins and employ a site labourer/cleaner to keep the place nice and tidy.

You do not want people eating and drinking in plantrooms where the site boxes are. Personnel have to eat and rest. If you do not provide this facility, then they will find a place on site to do it. This will create rubbish accumulation problems, and left over food will attract rats, which introduces the risk of Weil's disease.

If smoking is not allowed on the estate, designate the area outside of the estate where smokers can smoke. Conduct a toolbox talk on the estate's policy and show the smoking area on a map during the site induction.

Provide toilet facilities with hot tap water and soap unless the client allows you to use his toilets.

Provide a drying room. The drying room should have storage cubicles or pegs where guys can change their clothes and provide adequate heating so that working clothes, like overalls, can be hung to dry overnight. If you do not provide changing room facilities, people will be changing in the plant rooms near the site boxes. The problem here is that they will need to enter the plant room with full PPE. With no changing room on site, the operatives will simply be unable to do it. Clothes in the plant rooms also present a fire risk and are simply unhygienic.

Agree with your customer where to place a skip. Make sure that the skip you or your subcontractor supplies are lockable, otherwise other contractors who have nothing to do with your project will fill it with their rubbish. Tell me what you can do when you notice that the skip is filled with rubbish overnight that didn't come from your project? Are you going to dump it all out?

No matter what you write in the contracts with subcontractors, and no matter how convincingly they pitch that they will have "just in time delivery", there will be materials stored on site. Pipes, valves, plate heat exchangers, pumps, fans, fittings, lighting, lagging will appear here and there unless you provide a dedicated space for their storage. You also run the risk of being pulled by your customer for not protecting your equipment from theft and the weather elements. So provide storage: it could be a few cabins and/or just a fenced-off area.

Do not get in to long disputes with your subcontractors regarding their responsibilities to keep their areas clean. Take pictures and get your site labourer to clean it up. A clean site is a pleasant-to-the-eye site; it gives a perception that everything is managed well and is simply good practice. A clean site energises people; it is also great for keeping everyone safe. You will be highly respected by your client for having a clean site.

Prefabrication and off-site manufacturing

Some prefabrication will happen off site. This simply means that you cannot inspect the progress of works as regularly as you can on site. This presents you with a problem. From experience, you cannot trust your subcontractors when they say that pipework is being prefabricated off site. There are simply too many parties involved for things to go wrong or for the other job to take priority over your project. Normally, your subcontractors won't run prefabrication workshops: their subcontractors or suppliers run them and you have very little influence over them.

No influence means big problems.

There is only one sure way of being certain that the stuff really is being prefabricated and manufactured off site, and this is by visiting the manufacturer's workshop. Be careful that these manufacturers do not pull the wool over your eyes during these visits. Make sure that the equipment and prefabricated pieces are for your project. Check the pieces against the installation drawings and technical submittals. If necessary, initial the pieces with a permanent marker.

During pre-fabrication, pay attention to the gaps between the mechanical and BMS specs. Make sure you get all the correct features, power rating, BMS control interfaces etc. before the equipment is shipped to site.

Check out the electrical panels and the BMS control panels off site against the quality check lists. Make sure that labels are correct, lifting eyes are in place, panels are built to the design spec of the latest revision, that there are no defects, and switchgear and protective gear is up to spec. Run an auto-changeover test if applicable.

Run Factory Acceptance Tests (FATs) if the equipment is expensive and the project is big. For example, run the chiller against the load. Invite specialists from other key subcontractors to these tests in order to flush out any coordination issues.

Welding

Check evidence of welding inspections, such as welding inspection reports and welding maps. The welding inspector should be an independent qualified person. The principal contractor and subcontractor need to agree whom they should use. The welding inspector will make sure that welding is carried out as per specification. He will also make sure that the welders are coded for the specific method of welding specified in the specification (stick, TIG etc.). The coded welder will follow the procedure of the preparation of the pieces for the weld and will do the exact amount of weld runs around the joint as per welding spec. Beware of welders who cut corners. It is the welding inspector and supervisor's job to ensure the quality of welding.

Read the welding inspection reports from the welding inspector. All welds should be inspected visually. By looking at the welds visually, the welding inspector will be able to pick up faults. He can fail the weld on the spot or conduct additional tests like NDT (non destructive tests). Internal cracks on larger bore pipes can be identified with ultrasound. The welding inspector will mark the inspected welds with a permanent marker of some colour, like yellow. A lack of fusion will be a common fault on failed welds during the ultrasound. The weld, or at least a section of it, will have to be redone.

External cracks will be revealed by MPI (Magnetic Particle Inspection), which leaves white paint marks. This is how MPI is easily recognised. Use X-ray for Class 1 welding, which is normally used for steam. The welding inspector will ask each of the welders to perform a certain weld type (say butt weld). He will watch the welder doing it and then, normally, take the weld off site to X-ray it. You will see the X-ray results in the welding report.

You can also pressure test the pieces off site, flush them or dip in a chemical bath. This will accelerate the time on site between installation and commissioning.

Site induction – set the rules straight

Site induction is the time to set your rules and expectation straight. This specifically applies to H&S, hygiene, cleaning, and that the workforce is skilful enough for their tasks. The site induction is not a place to make friends, be soft or to have a sweet conversation with new starters. The guys attending induction will no doubt litter, bring food to the workplace, not wear the required PPE and may not be fully competent for the tasks they will be assigned to do. This brings huge risks to the principal contractor and contractors.

It is your responsibility as the principle contractor to keep the workers, building users, trespassers and visitors safe on your site. If something goes wrong, the question asked would be: "*Why did you fail to keep the lads on your site safe?*" If you are a principal contractor, it is always your fault. Use the previous sentence as the basis of your actions on H&S. Therefore, during the induction it is your duty to identify and filter through the people who are sound, know what they are doing and will do it safely.

Before site induction, ask the contractors to e-mail you copies of the following for each individual:

- ✓ Skills/CSCS card;
- ✓ Evidence of competences: welding certificates, NVQs, 17th Edition Wiring Regulations, Inspection and testing, Gas Safe, etc.;
- ✓ Any relevant tickets: PASMA, scaffolding inspection, MEWPs, Scissor Lifts, First Aider, Fire Marshal, etc.;
- ✓ CRB/DBS (criminal record checks) certificate;
- ✓ Asbestos awareness course certificate (for working in buildings built before year 2000).

Start with the CSCS cards. If they do not have a SKILLS/CSCS card, do not induct them. People undertaking unqualified tasks, like labourers and mates, should have at least a basic CSCS card or a certificate that they have passed the exam.

Once you are happy with their skills, make security checks for the individuals. The best way of achieving this is to have them send their basic CRB/DBS checks in advance of the induction. If the guys have access to children, demand enhanced DBS. An enhanced DBS takes up to 6 weeks, but it will have all the criminal records there starting from juvenile crimes and beyond. People may not want to share this information with you. In this case you have two options: the first is that you get your HR specialist trained to review the CRB/DBS certificates; the second option is to give guidance and rules to the subcontractor's management regarding the filtering process.

Shockingly, some 30-40% of your workforce will have some sort of police or criminal record. The enhanced DBS will show it. If it is something minor like a pub brawl that happened twenty years ago, you should induct that person. But with possession of Class A drugs, knife crime or robbery, do not induct these individuals.

At the end of the day, would you want someone like that doing a bathroom in your home? Why then would you let them into a customer's premises?

Create the site induction document in advance. The document should include general site rules, such as no people under the influence of drugs or alcohol; it should also specify the location of the toilets, canteen, drying room, changing room, smoking area and fire assembly point. The document should also have a map of the site with the above items identified. It should also explain what to do in emergency, the fire evacuation procedure, who the first aiders and fire marshals are and what to do if someone is injured or suffers a near miss.

After the induction, issue personnel with hi-vis vests and hard hats with your company's logo. This is to differentiate people who work on your project from the maintenance team and people working on the other projects in your area. You can also issue a sticker showing they were inducted on a particular day; the sticker can be adhered to their hard hats.

Stress that following PPE is compulsory: hard hats, safety boots and hi-vis vests. Additional PPE is as per task in hand and RAMS specific. Additional PPE may include Level 1 protection gloves, earplugs, face mask, gauntlets, safety glasses or goggles. Only their specific RAMS can dictate which particular PPE is needed. Stress that if PPE is not worn, worn incorrectly or misused, the person will be yellow carded for the first time and then red carded if caught doing the same again.

More importantly, follow it up with your actions. As difficult as it may be, you must remember that it is better to be safe than sorry, or worse, finding yourself jailed for allowing unsafe works to take place on your site.

Prior to allowing anyone to start any works, make sure you have task and job specific RAMS. Always review them personally, because you above all should understand what needs to be done and how to do it safely. The subcontractor's supervisor or manager should walk the areas where they are planning to start works, along with the guys who are doing the job, to assess the risks of getting the job done and produce site and job specific RAMS.

You also need to see a register of names for RAMS showing signatures that the operatives have read and understood the purpose and content of RAMS. In fact, do not induct anyone until you have approved the RAMS and have a signed copy of the RAMS by inductees.

All operatives working in buildings built before year 2000 must have undertaken a compulsory asbestos awareness course. This is required by law and includes anyone who will be in the building, including consultants and managers.

Dealing with accidents. Witnessing statements, investigations, actions and toolbox talks

Despite all your efforts and hard work, accidents and incidents will happen on site. You plan the works, closely supervise the works, thoroughly review risk assessments and method statements, routinely and rigorously enforce H&S on site, conduct H&S inspections, engage with individuals and listen to their H&S concerns. Yet, you cannot fully prepare for accidents and incidents to happen: you can only react to them.

When an accident occurs, act with urgency. Drop everything else you are doing.

Your first task should be making sure that the injured person receives medical assistance. If this is something minor, like a minor cut, your first aider might be able to treat it on site. If it is more serious, get the person to the hospital or call an ambulance.

You also need to make sure that no one else is injured.

Your next priority, and this is often forgotten, should be eliminating the risk that caused the accident. For example, if someone injured himself because of a hole in the floor, cordon off the hole immediately and put up a warning sign. Next, get the hole covered properly. This needs to be something proper and permanent, like a chequer plate.

When the injured person feels able to, make a record in the site accident book. The injured person or the person in control of the accident book can make the record entry. The accident book should be kept locked for data protection purposes.

The accident and incident could qualify as a RIDDOR incident (Reporting of Injuries, Diseases and Dangerous Occurrences Regulations 2013). There are a whole list of accidents and incidents that qualify as reportable under RIDDOR. The regulations have softened recently (2013), which means that should the injured person return to his normal duties and tasks in 7 days or less (including holidays and weekends), then RIDDOR will not be required. In other words, if the person is back to work in less than 7 calendar days, then there is no need to report the accident to HSE.

It may be difficult getting to the bottom of what happened during the accident or incident because some people will be afraid of getting themselves or others into trouble because of what happened. They will try to play it down. Your job is to get to the bottom of it.

Take witness statements face-to-face, record the time of the accident, visit the place of the accident, take plenty of pictures or record a video with facts. Get your H&S advisor to help you to compile the investigation report. Review the risk assessment and method statements.

In most cases, either the RAMS were not specific enough, or the injured person deviated from the RAMS. In any case, it is your fault as a principal contractor. If the risk was not highlighted in the risk assessment that your subcontractor submitted to you and which was approved by you, then it is your direct fault. You should know the environment where works will be taking place and the associated risks. You should also understand the works to be carried out so you can highlight any associated risks in carrying out the works.

In some instances, you might find that the workforce or the injured person has not signed the RAMS. Again, this is your fault by allowing works to commence or continue. To avoid situations like this, do not allow any works on site until you have a copy of RAMS signed by all parties involved in carrying out the task.

The problem is that you might have no experience with some specialist works. In this case, you need help of an H&S advisor or a colleague with experience in the specialist field.

The main thing about accidents is what actions you take to get H&S to the next level. Actions could be revising the RAMS for a similar task and getting registers with copies proving that every worker involved has understood it. It might be the toolbox talk you require to do to enforce the risks and show mitigating measures.

The problem is that until you experience some accidents on your sites, you will never understand your full responsibility. Having a safe site is the equivalent of a part-time job, but you can end up spending too much time resolving site issues, too much time with paperwork, and too much time satisfying the requirements of your client. With all these distractions, HSE is often neglected until you get a sharp reminder from your H&S advisor,

To get good traction, my advice is to undertake two H&S audits a week: one with your H&S advisor, and the other with the client's H&S advisor or CDMc.

The best set up is if you and your H&S advisor spend a few days together setting up the H&S system and files on site. The files will include RAMS, permits to work, track sheet for issued PPE, toolbox talks, induction, competency and training records including CSCS and skills cards, registers of means of access and harnesses, near misses, waste management file, including delivery and collection, weekly H&S reports file and weekly means of access inspection file.

Balance between site and off site work

You should aim to strike a balance between working from home and your presence on site. If you spend the majority of your time off site, you lose touch with what is going on. If you are the type of person who does not like to get bogged down in daily site issues, you will be spending most of your time off site. If this is the case, you must have a good team in place on site to deal with the daily problems that the site brings and for firefighting.

Firefighting will inevitably become synonymous with site management. This is specifically true during shutdowns and dealing with the consequences of them not going exactly to plan. The weaker, less skilled and less experienced your subcontractors are, the more chaotic, unproductive, stressful, nerve racking and frustrating your job will become.

On the other hand, you cannot spend 100% of your time on site. You will be spending way too much time reacting rather than planning, scheduling, programming, dealing with customer requests, reporting and completing paperwork associated with the project. In my opinion, you should be aiming to spend one day a week off site and working from home dealing with the administrative work. Of course, have cover on site during that day. Do not work from the office, as it is unbelievably unproductive; however, schedule all meetings on site.

There is another solution that many overworked and stressed out managers do: they put the hours in doing both the site work and project management. They come in at 07.00 and leave at 18.00; normal site activities run from 08.00 till 16.00. So this person, for example, handles e-mails in the morning and actions from 16.00 till 18.00. This is a sure way of developing mental health problems.

Another problem with such an approach is that, because no normal person works outside of normal working hours, you will never be able to obtain information, check something or delegate something to someone. These extra hours become an unproductive waste of time.

I personally know a project manager who was signed off with work related stress for six months and never came back to work after that. I know another one, who was signed off with the same diagnosis for three months. He came back, but took a lower engineering role. There is nothing heroic and brave in "putting the hours in".

I, myself, went through the stages of a get-it-done-at-any-cost mentality. I discovered my first grey hair at twenty-nine during a project I was working on. We were under enormous pressure because of the huge weekly £100K losses and we had to get the project over the line very quickly with very unqualified resources at our disposal. I did not crack, mostly due to my stubbornness and sheer determination of getting it over the line and by not giving up and quitting.

A sports background helped develop my mental strength skills to get through that project, but my family life was affected and my family members silently suffered. I write it now and feel guilty about it. I pray they excuse me for ignoring what is now the most important thing for me, which is my family.

My wife noticed that I could not switch off at home and was constantly under the heavy load of pressure from the client. When I was at home, my mind was at work.

Well, lessons had to be learned. What I lacked in skills, I compensated in stubbornness, sheer determination and discipline. You can probably get away with putting the hours in at the very early stages of your career until you get the necessary skills, but take my advice in the long-term.

The bigger the project, the more complex it is, the more time you will have to be spending recharging your batteries. Stress accumulates, so you should make a conscious effort to change the environment by working off site at least once a week. And because you won't need to spend hours commuting to the site, you can substitute these precious hours with sports, hobbies and family life. You have to maintain your physical and mental health throughout the project and life.

All successful project and site managers I know exercise at least twice a week. Sports where you are being constantly challenged by other team players, competition or a coach, in my experience, switch you off from work the most effectively. You have to feel good about yourself and be releasing the hormones of happiness by doing what takes your mind off work. Sports and hobbies are the best way of achieving this.

Biggest showstoppers during the installation
– the baker's dozen

1. No builders work completed

You can plan for the obvious builders works, however there will always be last minute issues requiring builders work. You can plan for the plinths and holes in the walls, ceiling and floor, but there will be a whole bunch of things you as principal contractor or contractor will have to cash out for. These could be things related to H&S and the wellbeing of the site's workforce: you might need to box in skylights to prevent people and materials falling through them; you might need to install chequer plates to cover open sump holes. Building temporary partition walls is not a rare thing either.

You might need to repair walls, ceiling, floor, change carpets and plant room doors. Your subcontractors might categorically refuse that they damaged anything. Aren't you the one in control of the site's H&S? Get the builders in fast to eliminate or substantially reduce the risk of a person's harm or a dangerous occurrence. You will get your subcontractors in the long run, but if the customer requests that you do something, or there is a risk through H&S, you will have to act fast.

Incompetent subcontractors do not plan for works promptly. This means that they face issues and act on them in a reactive way. This might mean a Friday afternoon email from one of them confirming that they cannot proceed with works until the walls are made good in order that they can hang the radiators. I strongly suggest you have a building contractor on hand to react fast. If you try to instruct the subcontractor to sort the problem out, they might first of all refuse the instruction, as it is not their type of work; secondly, the contractor will drag his heels completing the builder's works. He is overwhelmed with the other works, understaffed and frankly not too bothered about helping you out.

2. No power

Have a power-on request form. Anyone working with the mechanical contractor and in need of power will have to complete the form and wait until you sort out the power. During the construction and commissioning phase, power will be required for:

1. Welding sets (3 phase);
2. Flushing of the circuits;
3. Commissioning of the pumps, AHUs, fans, chillers, plate heat exchangers (PHEs), water softeners, pressurisation units, packaged equipment, heat and steam meters.

To kick off welding on site, you might need to provide a few 32A commando sockets with RCD protection. If you missed it in your contract, the mechanical contractor will request that you provide power. Finding where to get the power from and arranging the shutdown to connect to the existing board will become your problem.

If you do not plan thoroughly, the mechanical contractor will be delayed with flushing, balancing and commissioning. The key message here is for you to stay in control of the BMS and mechanical package. Review your program regularly and ask the question: "When will the power be required?" The second question for the BMS/electrical contractor is: "*How soon will power be provided?*"

Before you can energise the pumps and fans, remember that the BMS design (points list, description of operation and wiring diagram) needs to be submitted and reviewed. The panel needs to be built and tested. Then it needs to be shipped to site. The plinth for the panel needs to be built. Then the panel needs to be moved into position and probably using a plant movement contractor. If the panel is delivered in sections, it has to be bolted together and tested again in its final position.

Before you can provide power, you also need construction-issue installation coordination drawings. This is to make sure that cable trays and trunking does not clash with the pipe work. Electrical installation needs to be completed and tested, phase direction for the pumps and fans to be checked. Only then can your client's AP (Appointed Person) energise any equipment.

As you can see, there are lots of hoops to jump through before you can provide power for flushing and commissioning. You need to become an expert in the management of the BMS package in order to succeed with the overall project.

3. Can't shut the mechanical or electrical services down

If you cannot shut the heating, steam, condensate, DCW, DHW, AHUs or electrical distribution board, this can be a real bummer. In an existing building, the shutdown needs proper arrangements, planning, notice and coordination. Oftentimes, the services cannot be shut at all. You might need to wait until the end of the heating season or the next planned maintenance shutdown to complete your works.

This can seriously screw up your program.

To move forward, you might need to hire some temporary boilers. They could be electrical or oil. You might need to run temporary pipework and even arrange a short shutdown on the system to connect your temporary hoses. You might consider using one of the standby heat sources from the other system to temporarily provide heating or DHW for your system.

The department affected with your work might need a two-week notice so they have enough time to arrange an alternative space for the people to work from. There is nothing you can do about it. There are some production processes or places, like operating theatres, where you might need to wait for months for the shutdowns to happen.

There is only one piece of advice for scenarios like these: utilise the past experience of the maintenance team to understand which areas will be affected and how easily shutdowns can be arranged.

4. Wrong materials

I bet you are thinking that I am insulting your intelligence by bringing it up. Dear reader, I only speak from personal experience. You spent weeks going back and forth with specifications and technical submittals; finally, you now have it all sorted and approved. You think you can relax now, but you

realise, to your amazement, that non-conforming materials, accessories and equipment are being installed.

Normally, this happens under the pressure and rush of starting the installation; it also happens when you have already stripped out the equipment and the weather suddenly changes for the worse and the building urgently needs heating. Complaints are quickly escalated to the director of estate or even the board and you end up executing a quick-fix installation just to get the heating on.

In the above instance, a customer requesting heating cannot be ignored. The smart thing to do is to avoid stripping out until you have everything in place to do the work.

This type of mistake can also come from basic miscommunication where the installation supervisor, who, of course, never attended the design meetings, has an old version of the specification or technical submittal. Even worse happens when the installation manager or supervisor does not bother reading the spec and technical submittals and places the order based on the most commonly-used industry items. It might also happen that the correct materials are on a very long lead-time. The pressure to get the system on might push the contractor to use materials that are not up to spec.

5.   No installation drawings

The design engineers might still be arguing about the design issues, and you still do not have installation drawings. You have a bored site crew with no work; the guys get frustrated; you get frustrated: there is no progress. It might be the case that there was a design change and a variation is required.

There is another option here that happens even more often: the installation team proceeds without the construction issue installation drawings. The consequences are obvious. Parts of the works will have to be redone. This is a real bummer, because re-doing something takes twice as much time and money. These types of error, this early on, will hurt your timing badly. Months later, you can run out of cash and you will face a grilling from your superiors. Initial mistakes add up and eat up profit margin.

The advice here is to finish the design first and then start on site.

6.   No free issue items from third party contractors

Although the contractors are down to coordinate their works, this is not normally the way things are. The contractors look for any reasons outside of their remit to justify and blame something outside of their contract: "*We have not installed the panel because the plinth is not installed.*" If you do not really know what's happening, you will buy into their deception. The reality might be that the panel is in the workshop weeks away from completion because not all of the components arrived to the off-site workshop.

Here is the list of the free issue items to hunt down:

- BMS pockets, DP switches, pressure sensors, flow switches;
- Steam, heat, electric and water meters;
- Control valves;
- Large and expensive equipment such as chillers, AHUs, fans, FCUs, plate heat exchangers, water softeners, pressurisation units;

- Luminaries and lighting tubes;
- Electrical and BMS panels—this is so the BMS installation company can just install and wire the electrical services and BMS control.

7.   No access

You will often struggle with access to some areas, and you could find that the area is occupied at all times. What can also happen with older multi-department buildings is that each room has its own key or code. The funny thing is that security does not often have the access to some rooms. Yet when you plan the works and notify everyone to come and do the work, the rooms remain locked. Your simple job of putting on TRVs becomes frustrating and impossible. Instead of cracking on, your primary job is overtaken by the more difficult job of negotiating access with the building's users.

Similar issues with access arise when you need to trace some existing pipes or have access to the services above the ceiling tiles or worse: above a solid suspended ceiling. This happens all the time. In this situation, you are stuffed. Alternative routes for these services might be required.

8.   *Existing isolation valves not holding and plantroom gullies leak*

You plan to isolate part of the system, come on site during the weekend to do the job, only to find that the existing isolation valves are not holding. To everyone's disappointment, you cancel the whole thing. You or the maintenance team's first task becomes replacing the faulty isolation valves. For that, of course, another shutdown needs to be arranged to freeze the section of pipe or even drain the whole system . . . and that is "lots of fun" as you might guess. To make matters worse, the plant room you are in might not have a functional or adequate drain system in place.

The other issue with the draining down is that after refilling, the system will have to be vented and this might mean venting individual radiators. You cannot program your project accurately unless you deliberately plan for these hiccups to happen.

9.   Asbestos

The presence of Asbestos Containing Materials (ACMs) is the biggest stopper of a refurbishment project's progress. There is much fuss about asbestos, yet there is not enough clear and concise information regarding its exact location. Asbestos management survey documentation is unbelievably unclear; even H&S advisers can't make sense of it. References of the locations could be as clear as mud.

My advice is to get your customer to spend money on a demolition survey for each of the areas in which you will be working. Every location containing ACMs has to be fully cleared. I will expand on why this is.

You cannot possibly supervise multiple contractors in a plant room containing asbestos; it's physically impossible, so my advice is not to put yourself in that position.

For example, let's say the demolition survey identified ACM residue on the pipe work under the lagging and some residue on the ceiling in the plantroom. Although it is safe to continue work, it becomes a serious incident if your installers tear out a small piece of the old lagging to strip out the calorifier, or if they drill through the ceiling to bring pipes in. In both instances they disturb asbestos that becomes airborne.

The asbestos removal company will not remove asbestos-containing fuse carriers, as they are not qualified electrically to do so. This is a problem when the old distribution board (DB) needs to be stripped out. The solution, first of all, is that the electrical DB has to be electrically isolated and made safe; once done, the asbestos removal company will clean the board of the ACMs and dispose of it in accordance with the regulations.

Another area where the asbestos removal specialist will leave alone will be flanges containing asbestos. Depending on the age of the system, you may well find that all the flanges have asbestos gaskets. In this instance, all the specialist will do is place stickers on those flanges. Your mechanical specialists will then have to cut the pipework on both sides of the flange, bag the removed section, put a sticker on it, store it securely and then get an asbestos removal specialist to remove it and dispose of it as per regulations.

If there is a suspicion that asbestos was disturbed, a quick and easy way of finding out is to test the air, which, fingers crossed, will show a satisfactory result. Naturally, there are asbestos fibres in the air. So, a satisfactory result means that the air has fewer asbestos fibres in the air than the maximum threshold. All you will normally see in the air test certificate is that the result of the test is satisfactory. It means it is safe to work in the area.

Unless you fancy ending your career and possibly spending some time in prison, the only sensible way to safely proceed is to remove ACMs before allowing your team in.

Who said refurbishment and retrofit projects were easy? Do you not need to deal with asbestos in new-builds?

10. Can't flush the systems

Although this is the proper way of doing it, I do not advise flushing the original LTHW or CHW systems entirely. If the systems are old, you will cause no end of new problems for the estate team and building's users. Whether you or your client pay for them, specialist contractors will need to be involved in removing the sludge, rust and dirt accumulated in the existing systems.

Flush only the bits that you have installed. To do that, make sure that you have installed flushing loops. Because you do not flush through the plate heat exchangers, install a flushing loop or valves near the PHE. The distribution headers might also need to be connected to complete the flushing loop. Make sure you have installed stubbings (T-pieces with isolation valves, also called 'tap offs') with the isolation valves.

Do not assume that you can use CWS for the raw flush. Sometimes you need to find water from the mains, because water from the tanks can have a coffee colour. Do not be surprised to find out: some water tanks are not cleaned regularly; some aren't cleaned at all.

11. No competent labour

It is a sad case if you do not have competent labour. You will have to be a maverick to pull that project through. Long live mavericks! However, the construction site is not a battlefield and heroes are not appreciated here, nor am I asking you to be one. You will have all sorts of unpredictable and unusual problems if the labour on site is not experienced enough with the tasks they are doing. The worst cases will involve incidents and accidents. People will hurt themselves doing something they are not competent enough to do.

For example, it is no good having a crew specialising in new-builds on refurbishment projects. There are certain tasks and limitations with the existing systems that will trip the most experienced pipe fitter or electrician who has not worked with existing, functional systems before. The isolation of systems will be one of them, as the location of isolation points will become a nightmare over time. Old buildings were built at a time preceding current building regulations. Having 17th Edition Wiring Regulations training with no experience of working on old electrical infrastructure is a health hazard!

Quality of work, or more specifically, the absence of it, will be a constant issue that can only escalate when using an incompetent workforce. In fact, everything will be a problem for you and your customer. Respect yourself, your team and your customer. Get quality supervision and labour.

Your standard of work and H&S is not supposed to drop if you have incompetent labour on site.

There is only one solution: use contractors that you trust and who performed on very similar projects for you in the past.

12. Historic problems with existing system

The list of potential problems with existing systems is enormous. They will get you into serious trouble if you're not careful. Here are just a few potential problems:

- ⊗ No balanced heating or DHW systems. This is where some parts of the building do not get enough heat or DHW is not hot enough;
- ⊗ The water used for heating and DHW contains rust, plaque, scale and sludge;
- ⊗ Electrolytic corrosion does not just eat pipes away: over decades, electrolytic corrosion and the huge deposits of scale can halve the bore and the flow rate of the pipe. This happens if you have galvanised steel with copper on DHW or DCW;
- ⊗ No one knows which parts of the building the electrical or heating circuits feed;
- ⊗ Systems like heating, DHW and AHUs have had all sorts of historic problems ever since they were built and commissioned: pressurisation units keep locking out, BMS keeps locking systems, complaints always pour in from the building's existing users and circuit breakers keep tripping;
- ⊗ LTHW, CWS and DHW circuit crossover—this happens with old systems that were extended on numerous occasions;
- ⊗ Safety valves keep operating and flooding the plantrooms.

Once you start modifying the existing systems in any way, the historic problems of the systems inherently become the problems of your project. The only way out of this mess is to complete proper validation (dilapidation) surveys prior to starting the project. Your aim during the validation survey is to acquire knowledge of the systems and record it.

This will involve interviewing the knowledgeable personnel of the maintenance team and local contractors who constantly carry out small projects on those systems. There are always four or five preferred contractors who receive constant work from the client. These contractors are normally a very small operation; however, they possess the critical knowledge you need in order to make the systems work.

The trouble is that the small contractors doing these jobs are awful when it comes to recording anything they do (hence, one of the reasons why they are cheap). In fact, nothing gets recorded. Basic things, like schematics, are never updated.

The only known method I know of gathering the required information is by giving project work to these contractors. These contractors are normally very good at doing the work. However, a word of caution: add a line in your preliminaries for H&S and O&Ms management.

13. Leaks on existing pipes

Older buildings with the heating, steam, condensate, DHW and DCW systems operating beyond their life are full of leaks. Leaks don't just cost water losses, they are a burden on your energy spend. Due to leaks, softened, cleaned and de-aerated water is replaced by fresh, hard, acidic and untreated cold water, which will corrode your system and will causes the further formation of plaque. The pressurisation units will be constantly operating trying to maintain pressure in systems that are full of holes—it's like trying to blow up a balloon with a hole in it. The pressurisation unit will be burning itself out through constant operation.

All leaks need to be fixed before you operate your refurbished system, otherwise it will corrode your newly installed equipment and cause low-pressure problems, which will lock out the pumps for protection. If you install water treatment equipment, leaks will burden it with non-stop operation. Your warranties will also be affected for the boilers and plate heat exchangers. Furthermore, chemical dosing will need to be more frequent to keep the PH levels right.

The way to deal with leaks is via a maintenance work orders schedule. You raise a maintenance ticket requesting that the leak be fixed. You write a good description explaining where exactly it is and what is leaking, e.g. valve, flange, screwed connection or a pipe. You attach a picture. The client's project manager should approve this ticket, so the helpdesk assigns the operative or engineer to repair it.

Although the leak repairing works are managed by your customer, you should monitor it closely . . . it is a common experience that maintenance engineers turn up, don't have the right tools or materials to do the job, walk away from it and then close the ticket and book three hours for it. The only way of monitoring these kinds of issues is by maintaining the schedule of work orders weekly and keeping track of the progress.

It is often the case that, to repair the leak, the systems need to be shutdown. Although, your client should arrange these shutdowns, it is simply too much hassle for the underpaid maintenance engineers. Unmanaged and unplanned shut downs cause complaints. Be persistent and offer help in making sure that those leak repairs take place; otherwise, you will be staying on the job till the end of the heating season waiting for the leak to be sorted just so you can install some lagging or a few valve jackets. Multiple leaks add up and increase the time you'll be required on site. This will burn through your prelims faster than you think.

Managing the customer

Unless you bribe, dine and wine the customer, your relationships with the customer will be a roller coaster with ups and downs. This is inevitable. Although it might not be obvious from your customer's behaviour, your customer is also responsible for delivering the project. There will be quite a few senior stakeholders and people involved in the customer's team. Each will have their own agenda and want specific things from the project. Generally, this means conflicting directions and more work for you.

Your customer will recruit a team to help to run the project and provide technical support. For those outsourced project managers, consultants, cost consultants and H&S advisors, their customer is the one who gives them work and pays their invoices. They work for their client.

Often, independent project managers and consultants have conflicting interests. They need to deliver the service and keep the customer happy; however, the better they do their job, the sooner they will finish and will need another job. Having a project completed swiftly might simply mean being out of work sooner and then struggle to find the money for the mortgage.

Despite all the nodding during the walk around with you, the major risks introduced by the customer never quite make it in to the report sent to the outsourced H&S advisors. The reports are one sided with ALL the actions against your name.

Lots of the time you will be spending dealing with these political issues, so know the environment you operate in. Understand the organisation, its key influencers and underwater currents. The customer's internal politics will work in your favour if you harness its energy and get it right. Deal with the key decision maker face-to-face or by phone. Never bring up contentious issues in the progress meetings: resolve these issues in the background.

Aim to satisfy your customer by being lazy and with minimal actions for you. However, have the skill of saying "No" in a non-confrontational way. It is a skill that great project managers have developed over the years. They find great reasons why they should not do it and point out how it is in line with the customer's agenda.

You either lead the customer or you will be lead. It means having a strong focus on delivering the project and maintaining your profit margin throughout the project.

You need strong commercial support. By that I do not mean a bean counter, sorry, quantity surveyor. The bean counter will simply ask you to write a variation or a site instruction so he can put it on the headed paper and issue it to the subcontractors. No, you need decent commercial advice and support.

The commercial adviser will clarify when you can go to the customer for a variation. He will decide which requests for site instructions are valid and will need to be paid. The quality commercial manager will talk to awkward subcontractors. He will explain why they should stop threatening to stop the works and focus on getting on with the job. He will also explain to you which instructions are no-cost instructions. It is very easy to be dragged into site issues and paperwork. You have to dedicate space in your diary to go through the project review with the commercial manager and your line manager.

During the project reviews you should also review the risks to the project and often make difficult decisions regarding maintaining the progress at the same project margin.

Try to channel all the multiple customers' requests through customers' requests through the Estates and Facilities project manager. project manager. This will limit the amount of urgent and often conflicting requests from the key stakeholders of the customer's team. Ideally, the questions from the maintenance team also should be going through a single point of contact. The client's project manager will also be able to filter the rightful ones from those that have nothing to do with the project.

What are meetings for?

Meetings are not called to expand your social life: join a club for that or socialise after work. The meetings you attend are not to report and record the problems you are facing in delivering the project. The meetings are not there so you keep your customer up to speed with project developments.

The meetings should only be called to resolve a couple of specific problems.

The fact is if you need to have a meeting, then someone is not doing his or her job properly. That is the sad thing about it.

The show should be running so that you do not need to get heavily involved with it; if it does not, then you have poor contractors or designers working for you. In a perfect situation, everything should be running smoothly, without your involvement and on full autopilot.

Does that mean that customer and contractors will be thinking that you are lazy, that you are not getting involved and not pulling your weight? Will you withstand that pressure from them? What about the contribution to the project? How about the fulfilment of hard work?

It is not a thesis on management. If everything ticks along nicely, you are rarely required. This statement comes from the experience of successfully progressing works.

Your subcontractors, designers, suppliers and your client need to understand that you are not there just to be there – you see, some people believe that the project manager is supposed to be overworked, stressed out and always there when the phone rings in his on-site cabin.

As I said, if you have made the right choices with the subcontractors and the right decisions, the job will be running without you. You will be using your time focusing on the upcoming elements of work and relationships.

If you see that things are not progressing as they should and the situation is not being resolved, then there are malfunctions in your system. Address the cause of the problem. Resolving the issue itself will land the next problem on your table. Your project will become dependent on you.

If you are the smartest guy in your team, it is time to change your team.

Retire yourself from the mediating role between the contractors and opposing political forces in the customer's organisation. They need to come along to the table on their own without your nagging. I do not believe in mediation.

The meetings should be short. Keep them under 20 minutes. My favourite duration is 10 minutes. The best place for meetings is the plant room concerned. If there are design issues, then have the meetings take place in the plantroom or with A1/A0 layout drawings on a table.

Involve only the decision makers or key technical experts into the meetings. The people involved should be resourceful. No "naysayers" in the meeting and ideally on your project.

In the worst-case scenario with no progress being made, you will be having daily meetings. This is called micro-management and it does not work; rather, spend the time arranging for the replacement of the troubled contractor or their subcontractor. One of them is letting the project down. Don't tolerate that nonsense. Replace the failing party without hesitation or strengthen the team by replacing a key person.

If the package is performing, you might not need any meetings at all. This contractor may have a great team on site, including a fantastic supervisor. They crack on. They come to you from time to time. You go there just to continue building relationships. You get the occasional email from them asking to help to resolve some issue they face. Most issues will involve the client agreeing some changes or arranging a formal shutdown. You see, they have already spoken to everyone and all they are after is a tick in the box. You resolve it and do not hear from them for another two days. Every time you walk the job, you see their progress: the contractor's guys are happy doing what they like, and the contractor's supervisor never brings you problems.

Your client might want weekly meetings. Now, that you are getting all the weekly documents from your subcontractors (see next chapter) all you need to do is to collate them in a master document containing all the packages. You will be updating those for your customer weekly. You issue them to your customer and they can see where you are straight away.

Your customer might also want to record those meetings formally. If it is their meeting, they should record the minutes, not you. If it is your meeting, you can bring an administrator with you. However, remember that the actions from the meeting are for you to act on, not to record.

PART 4

Special bonus feature. Guaranteed system for managing your project successfully

Guaranteed system for managing your projects

I am going to share with you a system of managing multiple packages of subcontractors on site. If you follow the system, you will know exactly where you are at any point in time, which is a very powerful position to be in. I call it a position of control, power, and it comes with a very satisfying feeling. This system of work should be ingrained in the contracts with the subcontractors. It also needs to be made clear to the subcontractors' managers that these are the rules of the game.

Aim to be a true master of project management. To really understand where you are, have ten magic strings attached to each major package (subcontractor). To get the performers back on track you will be able to gracefully pull the required strings.

Normally, you will have multiple contractors installing different elements of work: this could be lighting, heating, pipework insulation, district heating, BMS, cooling, AHUs with ventilation and a building contractor.

Most people use the Microsoft Project programme with a drop line. It is an extremely useful and powerful tool to see where you are; this helps to make big strategic decisions influencing the progress. However, if the contractor is not performing and missing dates (and oh boy does that ever happen), you have to dig deeper and find out what the real problems are.

To really know where your project is, learn, understand and fully utilize the Holy Project Management Diagram given below.

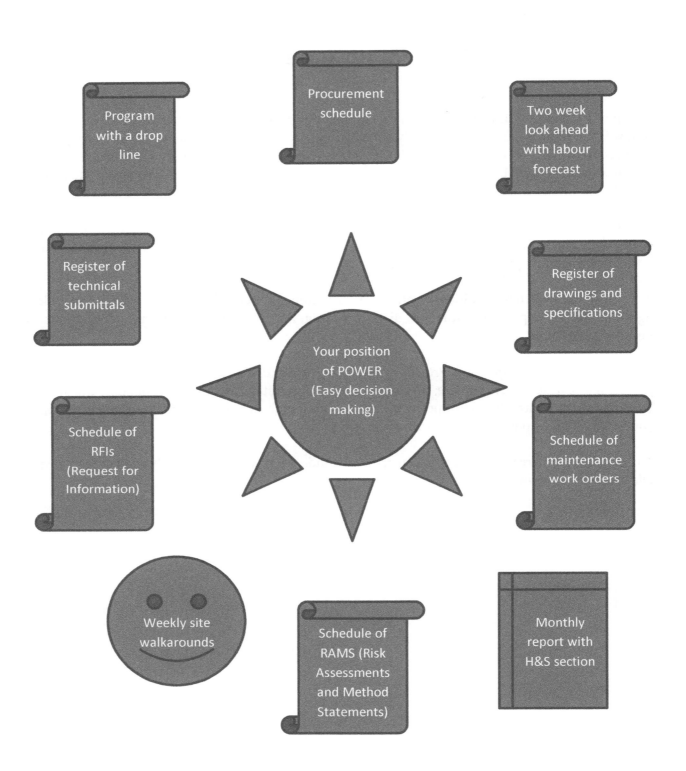

Stop taking the words of the subcontractors for granted and form your opinion based on the following items given in the Holy Site Management Diagram above. Let's list the building blocks again and expand them a little bit more.

1. Program update with a drop line highlighting specific progress stoppers for each activity;
2. Two-week look-ahead outlining labour plan and tasks required (to include tasks to be completed by the other contractors, i.e. power on, builders works, etc.);
3. Detailed procurement schedule;
4. Register of drawings and specifications;
5. Register of technical submissions;
6. Schedule of Request for Information (RFIs);
7. Schedule of Risk Assessment and Method Statements (RAMS);
8. Schedule of maintenance work orders (I will explain later how this comes into the equation);
9. Monthly progress report with a section on H&S;
10. Regular site walkarounds.

One essential requirement for this perfect system to work is that you have got to walk the whole site twice a week marking the progress against the project program. This will give you feedback on how the contractor is performing based on real life and not what is written on paper.

Let's touch base on each of these magical strings attached to each package via the contract:

Program update with a drop line highlighting specific progress stoppers for each activity

First of all, the program needs to be detailed enough. Breaking down the areas and elements of work does this. The drop line (shows exactly where you are with the each element of works) will show how far the elements of work are behind or, in rare cases, ahead of the program. This will highlight the packages with problems. This is the main point of the program: it tells you there is some sort of problem here needing more attention and investigation.

The program should also show the showstoppers, e.g. issues that, or are just about to, stall the progress of the package. These could be anything outside of contractor's scope, like builders works, powering equipment, asbestos in area, outstanding design issues, missing instructions, variations, unanswered RFIs, outstanding maintenance issues (such as leaks), coordination issues, operational constrains of the functional building (can't shut off the services) and free issue equipment provided by others that may not be on site.

All those listed items are now your problems to deal with, act upon and resolve urgently. The thing is that contractors have a disease called "excusitis"—they are great at blaming something or someone. They will argue that it is in their contracts that they coordinate with other contractors, that they've done their due diligence, and that you did not buy a turnkey solution.

The funny thing is that the commercial managers and quantity surveyors do not run the projects: you do. They spend too much time in their offices. Their function is back-office, so they have minimal impact on a project's success. They won't be getting their hands dirty with the project.

The reality and irony of the contracting is that contractors need to be managed. Period. Some of them are barely competent at what they do, are under huge pressure, are under resourced and, because of all of that, are stressed out. They will not even start contemplating going to the area and carrying on with works unless the issues they raise are resolved. This is the reality of it. They stop the progress in that area completely and pull an under resourced team to another area, which, by the way, is also behind schedule.

So, please get it. Always make sure that the contractors have all the showstoppers listed for each specific area as a starter. Secondly, and which is the most important part, always resolve these issues on the spot as a matter of priority over anything else you do.

In fact, sorting the problems brought up the subcontractors is the most critical thing determining the speed of the progress of works.

If you can balance sorting out the project problems with customer satisfaction, then you are an outstanding project manager.

So, pick up the phone, issue that instruction, get the builders on site, get to your customer's office and shake out the answer for that RFI, arrange an urgent coordination meeting between the BMS and mechanical contractor, hire your own labourer and clear that plant room from dirt and rubbish, visit the prefabrication workshop and incentivise weekend work and base the incentive on performance (get the job done and go home).

Do whatever it takes straight away, but get it sorted. If you do so, the subcontractors will understand that they cannot use these raised issues as an excuse for not completing the works they are contracted to complete.

Two-week look-ahead showing labour plan and tasks required – to be completed by the other contractors

This is one of the most important and underutilized tools. You have to share the two-week look-ahead with your customer and maintenance team. As they can see where you are working, they will be able to deal with complaints pouring in to the maintenance helpdesk; it also shows them when shutdowns to expected and, more importantly, have enough time to notify the affected departments. You revise the two-week look-ahead every week.

Here is the golden rule. The two-week look-ahead is a realistic, achievable and true representation of the works in the next two weeks. In other words, it is real.

It is an Excel spread sheet listing the tasks you are doing in each location also showing the dates when you are doing them. It also shows the labour assigned to each task. This tool also shows you how many people are on site and helps you to decide if you need more.

The two-week look-ahead must also show the activities required by the other contractors, client or anyone else in order to progress. As an example, it may show that power is required on Wednesday for the pumps 3A and 3B, or the plinth is required by next Friday. This tool is also good for your walkarounds, including H&S inspections (shows working areas).

The two-week look-ahead should be distributed to your client, maintenance, subcontractors and your project team.

Detailed procurement schedule

This document will have all the major items required for the installation to take place. As an example, it should have plate heat exchangers, boilers, water softeners, pumps, pressurization units, AHUs, chillers, LVDB (Low Voltage Distribution) panels, BMS panels, lighting fittings, control valves, accessories with a long time delivery, fans, attenuators, TRVs, VSDs, meters, BMS controllers, windows, step overs, steam valves, etc. It is very easy to create such a schedule based on the equipment schedules created during the design phase.

Once your subcontractor listed all the items required to complete the project, then based on the program dates, he can put a day for each of the items by which the items need to be on site. Working backwards, the suppliers need to be contacted to establish the estimated delivery time once the order with them is placed. It often takes 8 weeks from the order until the new plate heat exchangers arrive on site. The same goes for the complex water softening plant. It takes 6-9 months for CHP or larger plant to be manufactured as well.

The orders need to be placed well in advance so that the required equipment arrives on site before the installation commences. Otherwise, your workforce will not be able to progress effectively.

Another common fault is that it might be the case that your contracts with the subcontractors have not yet been signed. This is not good if they only operate on a letter of intent covering one-tenth of the contract value.

Please do not be naïve: your subcontractors will not place any orders until they get hold of the full order or, even worse, until the contracts are signed by the directors (the contract negotiations can take months for larger projects). In this case, realise that you are holding yourself up; the contractors are not to blame. You will often need to look in the mirror before 'screwing the subbies'. It takes guts and leadership to accept the responsibility.

Register of drawings and specifications

Again, it is up to you to make sure that all drawings, schematics, specifications and schedules of equipment are the construction-issue revision before works start on site or, preferably, before you go to tender to select the contractors. It is silly to think that you can continue doing design development in parallel with installation. In fact, it is a recipe for disaster.

Every time you make a change, the contractors will rush to notify you about the design change and will demand a site instruction covering the associated costs involved. You can spend your entire budget and eat into your margin if you allow the installation to start before the design is fully signed off and is construction issue. The rule is simple: you design, and then you build.

The same applies to the Design and Build (D&B) contracts. In D&B, you select the contractors who design the job and then build it.

The register of drawings and specifications is a document listing all the documents produced. It has the dates of the issue and shows the type of issue, e.g. preliminary, construction or as built. The register will show you which documents have not been produced yet. As with the procurement schedule, the drawings and specification dates should tie up with the installation program. Drawings and specifications need to be approved at least two weeks before the works' commencement; this is to allow for mobilisation of labour and prefabrication off site.

If the design is being delayed, speak to the design managers, find out what the issues are and sort them out. The classic issues holding up the design are unanswered RFIs and incomplete dilapidation surveys.

The dilapidation survey will establish flow rates, pressure drops, pressure set points, running temperatures, lux levels, the currents of any running motors (for example, lifts, pumps and AHUs), power factor, harmonics survey and a survey of the electrical loads. The designers will also have to make sure that the isolation valves are operational and that the existing systems are functioning correctly. If you do not have this information, you cannot design the refurbishment of the systems.

Once you know what the real showstopper is with the design's progress, resolve the problems immediately.

Schedule of technical submittals

People often do not see the difference between the technical submittal and the specification. Some specs can actually tell you that the pumps should be, for example, Grundfos or equivalent. ; in fact, the choice of specs might not be as impartial as you'd like, as the design consultant might have particular preferences. Ideally, every single piece of equipment down to the screw needs to have a technical submittal.

The technical submittal is a data sheet with a description explaining exactly the type and model of equipment proposed. It must have a section on installation, operation and maintenance as well so that the maintenance team can comment on it.

The contractors must not be procuring the equipment without an approved technical submittal, and you should not allow them to do so.

Each contractor should create a schedule of technical submittals, updated weekly, and along with the other documents touched on in this chapter.

If the technical submittal is not yet issued, you know that the equipment has not yet been procured!

This document, along with the procurement schedule and programme, tells you that you have a problem. You can chase up the suppliers to get accurate data sheets or chase the contractors.

In some instances, where there is nothing similar in the market place, one may have to deviate from the specification to find the closest matching product. Of course, this needs to be agreed with the designers, client and often other contractors as well. Be aware, that any deviation from the specification can have commercial implications. Your client might ask for money back if you buy substandard items.

Schedule of RFIs

Every piece of information you require from your customer needs to be raised as a Request for Information (RFI). There will be tens of RFIs at the design stage and tens of RFIs during the installation. Word them in plain English so that people can understand your requirements. Before issuing the RFI, talk the question through with your customer. This is because many clients are often wary of the angle of some of RFIs, and often rightly so.

Some contractors will try to trip you or dodge their responsibility. Often, the RFI system is heavily abused. This is when commercial aspects, such as request for an instruction, are brought to the table. Another thing that often happens is that the contractor is too lazy to read the contractor's tender pack and asks the information already issued. Respond firmly that the system is not to be abused and reject inappropriate RFIs.

As my favourite business philosopher Jim Rohn used to say: "Walk the higher road".

Never pass on RFIs from your subcontractors to your client without reviewing them. The customer too can often play the arsey game, asking you to clarify the RFI to the point where you have answered it yourself. Customers do not like direct responsibility. They are also often in a position where they do not have the power required to make the decision and act on it. For example, the client's project manager cannot make a decision on his own without affecting the capital budget, maintenance or service disruption. Please understand this important point.

Update and Issue the RFI schedule every time you issue a new RFI, receive the answer on the RFI, close the RFI and then put dates for each of these so that it becomes a proper monitoring tool. Give your client seven working days to respond to an RFI.

Schedule of Risk Assessment and Method Statements (RAMs)

Another schedule, you might say, but aren't we all responsible for our H&S and the safety of people around us by law? It is actually a part of H&S requirements for you to have all RAMS reviewed before the start of the activity, but also regularly (at least once a year). Every time you have an accident or a serious near miss, request that your subcontractors resubmit their RAMS to make sure that nothing similar happens on site in future.

Your contractors need to create the schedule of RAMS and issue it to you every time a new RAMS is submitted or the original one modified. The programme will dictate RAMS submission dates. Ideally, you want RAMS a couple of weeks before works commence on site. This will give you enough time to comment on them and for the contractor to revise them.

Reviewing the RAMS, you may impose additional requirements on training, supervision, Personal Protective Equipment (PPE) or anything else you think will reduce the risk of a person being put on harm's way. You can impose pretty tough requirements as a principal contractor if you wish to; however, do not become a little H&S policeman. Work with your guys and listen to the personnel in order to have a safe site.

Monthly progress report with a section on H&S inspection

The H&S section of the monthly report, which by the way is a summary of the progress during the last month with pictures, must contain the following as a minimum:

- Records of weekly inspection of means of access (ladders, stepladders, scaffold towers etc.);
- Portable Appliance Test (PAT) register (test to be done every three months);

- Register of means of access on site (ladders, stepladders, scaffold towers etc.);
- Accidents and near misses;
- Risks introduced by the client;
- Evidence of weekly H&S inspections.

Schedule of maintenance work orders

You will have to complete and update this schedule regularly. This is something unique to the works in existing operational buildings, which is our retrofit.

The maintenance team will be heavily involved in systems shutdowns. It is very wise to create a schedule listing all the jobs (work orders) you have requested that they complete. This is so you can monitor whether the items are completed or not.

Let's say there is a LTHW leak on a pipe coupling and you are contracted to insulate this pipe. You identify the leak and report it to client's project manager and ask for the maintenance team to fix this leak. Other examples might be electrical or mechanical shutdown where the client needs to isolate the service, problems with the existing systems, the temporary removal of pigeon netting, the replacement of faulty isolation valves and many others. I also add the existing H&S issues to the maintenance work orders requests.

You record the request in the schedule and update it weekly, adding the approval's date, job reference numbers (the maintenance helpdesk will give you those) and when the job is completed. This will indicate how far the problems introduced by the customer are affecting your progress. The customer also appreciates your time and effort in reporting the existing issues on site. Because you are a third party acting like an observer, it is likely that these problems will be acted upon.

Site walkarounds

The importance of weekly walkarounds is enormous. How do you find out if the subcontractors exaggerate the progress in their documentation? If you walk around the job twice a week, no one will be able to pull the wool over your eyes. Walkarounds complete the loop of control. It is real-life feedback, if you want. You will have a full picture and can make tough decisions and take corrective actions. Soon, you will establish which contractors are performing and which are not performing at all.

With bad performers you might have thoughts on substituting some of their works with another contractor. In the worst cases, it might be a right replacement by another one.

By the way, to often resolve the problem you might be changing sub-subcontractors. You might even have no choice but to appoint a site or a project manager just to run this troubled package. The sooner you know about the problems, the easier it is to resolve them.

*   *   *

That's all about the Holy Project Management Process. This is the best system of managing the building services retrofit and refurbishment project known to me. As you can see, there are a lot of things to consider. It is essential to have control over all of 10 magic strings described earlier to have a successful project.

PART 5

Closing the project

Completing your first area: how to snag

It is very important to start handing over the areas and systems as soon as possible. Spreading yourself thin over the entire scope of works is a dangerous tactic. As soon as some areas are very close to completion, focus on those areas to fully finish them off. This means getting your installation ready for the handover, snagging and de-snagging the area, getting the as-built drawings right, tagging, placing valve charts and schematics on the walls and labelling the equipment, systems and parts. Completing the O&Ms and training for the maintenance team are also an essential part of the handover.

During the snagging, check the following:

General:

- ✓ Area is clean;
- ✓ Make sure that firestopping is done;
- ✓ Check the installation against the as-built or record drawings and schematics;
- ✓ Asset labels installed;
- ✓ Labelling of the services and equipment completed.

Electrical:

- ✓ Check cables, pipework, accessories, installation methods and materials against the approved and latest technical submittals;
- ✓ Check quality of installation (get help from the expert if you need help with some works);
- ✓ Check the containment, including conduits and trays;
- ✓ Review electrical testing certificates;
- ✓ Check panels: power and control sections;
- ✓ MCB reference charts are installed and up to date;
- ✓ As-built panel wiring diagrams enclosed in every panel and laminated on the wall.

Mechanical:

- ✓ All the heat, steam and water meters have been commissioned by the specialists and are fully functional on aM&T with the validated readings on the system and on the meter display;
- ✓ Every accessory, like valves, commissioning station and strainer, has a tag;
- ✓ There is adequate provision of pressure gauges, temperature gauges, air vents to be installed at each highest point, drain cocks to be installed at each lowest point;
- ✓ There are gradients on the pipes for draining;
- ✓ Every piece of equipment, like chiller, pump, plate exchanger and pressurisation unit, are labelled;
- ✓ Make sure that differential pressure switches and sensors are installed and piped up;
- ✓ There is the latest schematic displayed in the plantroom. There is a laminated valve chart on the wall;

- ✓ Make sure you have seen pressure testing, flushing/chlorination/balancing and commissioning documentation for each piece of equipment;
- ✓ Pipework is lagged as per specification and cladded;
- ✓ System identification labels are in place and flow direction is provided;
- ✓ Welding maps provided;
- ✓ System is tested against the load and is fully functional.

The snagging process should be as follows: get the de-snagging sheet from your subcontractor who is in charge of that particular system. This is to make sure that the contractor snagged and then de-snagged the works himself first. You also need all the commissioning documentation, pressure-test results, flushing results, electrical certificates, balancing sheets and as-built drawings in a pack. This pack can be a part of the O&M, which you can review electronically in advance.

Do not start snagging until all the items on the list are de-snagged by your subcontractor. Once it is all signed, you go and snag the system. You can invite your subcontractor on a walkaround to make sure that your snags are understood and acted upon straight away. You issue the comments with accompanied pictures and give a few weeks to your subcontractor to rectify the faults and snags.

Once you have re-inspected the snags and they are completed, you issue the whole pack to your client for inspection. Your client issues you with the snag list, you review it yourself and then with your subcontractor. After this, your subcontractor rectifies the legitimate snags and you negotiate with your subcontractor and your client regarding the ones that are preferential.

Mark your subcontractor's progress based on the snag-free installation and issuing all the handover paperwork including as-built drawings, O&Ms and documentation for the H&S file describing the H&S risks involved in maintaining and operating the systems installed.

I would never pay more than 85% of the project value until it is snag free with all the above mentioned paperwork issued and accepted by the client. Once you have all the paperwork, retain 5% of the project value for the first year to make sure that your subcontractor honours warranties and guarantees on the systems and parts.

Certification and commissioning

You should invite your client to witness the certification and commissioning. If the client does not turn up, carry on regardless. With complex witnessing, make sure that you are happy with what is being witnessed prior to offering it to your client.

Pressure testing

One important thing to note is that your welding inspector must inspect any on-site welding before the pressure test. Depending on the specification, this might involve visual inspections, MPI, ultrasound inspection and even X-ray. The condemned welds will have to be rectified and re-inspected before the pressure test.

Equipment like plate heat exchangers, existing pipework and water softeners are to be excluded from the test. You might need to spade them out (install a temporary metal plate to isolate water). The best way to separate the existing system from the new is by means of the isolation valve. It is great if there is an existing valve and it is holding. If not, replace it with a new one. If you cap off the new pipe, then you can't make sure that the demarcation joint is holding.

It is recommended that wet systems are tested at 1.5-2 times the working pressure of the system. If the working pressure of the system is 3 bar, then you test it at 4.5-6 bar as a minimum and sustain the pressure for a minimum of 2 hours. I recommend testing it at 2-3 times working pressure. You must use calibrated pressure gauges. The tested pressure (4.5 bar in our example) should stay constant for two hours without dropping. If the pressure drops even by 0.05 bar, then you have a leak that needs to be found and rectified. Once rectified, the test then will have to be re-done.

Flushing

You need to flush any new pipework that you have installed. However, I do not think you should undertake the flushing of any existing or remaining pipework. The existing system should have been maintained, cleaned, filtered and dosed. It is not really your job to flush existing pipework unless it is paid for as a variation.

The existing systems might have leaks, be crossed over with DHW, DCW, the other LTHW or even CHW system. This does happen with older systems. So, before you make any commitment, think twice. If you want to help your customer, get a specialist involved at an early stage. Besides, even if you clean the old pipework, it still needs to be maintained in future years, and if your customer was not maintaining pipework in the past, how do you know that this will be done in the future?

Employ a specialist to flush the new pipework for you. He will be able to advise you on the duration. He will flush the pipework with fresh water first, then he will flush it with chemicals. Once he is satisfied, he will take a sample of water and send it to the lab. You should get the results in about seven days.

Once the results from the lab show satisfactory results, it is down to your customer to confirm that it is okay to open the new pipework to the old system. As a precautionary measure, you can take samples of water in the existing systems and issue them to your client proving that the systems are full of iron oxide, dissolvent and show off-the-scale PH levels. If your new plate heat exchanger or boiler scales-up six months later, do not accept liability: your customer has not fulfilled his obligation to maintain the systems, which you should have specified and agreed in your contract.

Always make sure you also take samples of the water before the project started. It will help to work out the chemical content the refurbished system needs to be topped up to.

Chlorination

Chlorination applies to DHW (Domestic Hot Water) and DCW (Domestic Cold Water). As with flushing, you need to get a specialist on site to chlorinate the newly installed DHW. Chlorination will kill any germs and bacteria in the pipework and make it safe for people to wash their hands or take a shower. Short pieces of pipework can be dipped into the solution; for H&S purposes, however, make sure that the dipping tanks have labels identifying the chemicals and their associated risks. Once chlorinated, the pipework needs to be filled with DHW or DCW within seven days.

Balancing

You will be modifying the pipework, changing equipment and/or accessories. This will change the operation of the system, which might cause inadequate flow rates on some of the circuits leading to cold areas for the heating systems. To avoid this happening, I advise that you record the pressure drop and flow rates of the circuits before you undertake the stripping-out of existing systems or parts of it. If you have this information and provided double regulating valves for the circuits, then your balancing specialist will simply set up the system to where it was before you made the changes to the system.

This is the only way of making sure that your works do not introduce problems for the building's users. If you have not recorded the flow rates and pressure drops before starting works, then you inherit the problems that the maintenance team used to deal with. There is no way of protecting yourself. You touched the system; the problem is now entirely yours.

Commissioning

It is often forgotten that systems need commissioning. Many people do not understand what it even means. This is bad for the industry. In fact, I am sure it is one of the major reasons why the construction industry has a bad reputation. It is really sad. The worst offender is the BMS. The systems are set up and run in some sort of fashion. Most clients do not understand the intricacies of the functional operation of the BMS, which is why they stop even trying to get it right.

Commissioning is setting up newly installed systems so they run and function as per original design specifications and descriptions of operation to achieve maximum comfort for the building users and ease of operation for the facilities management team.

Ironically, there are more systems in the UK that are poorly commissioned than those that work properly.

BMS is not the only example: steam, electrical, heat, CHW and even cold-water meters more often than not clock inaccurate data. VSDs are another example where they are used for nothing more than a soft start and stop for fans and pumps. It really is a shame. Complex equipment with its individual control panels, like plate heat exchangers, water softening equipment, pumps, meters, pressurisation units, chillers, boilers, chiller sequencing panels, all need commissioning.

By commissioning, I mean obtaining a commissioning specialist from the manufacturing company itself and costs between £400 and £600 per day. Per day often means something like 4 hours including travel time. This is one of reason why you rarely see these guys on site, unless you demand it, of course. By the time commissioning's required, the money in your subcontractor's pocket gets very tight, and no one wants to pay for commissioning specialists. This is why many subcontractors will cut corners in the hope of getting away with it, unless you shout, of course.

Demand commissioning certificates from each of the specialists; also, attend and witness the commissioning. Always state in the contract that adequate time must be allocated for witnessing: you do not want to spend days with commissioning engineers trying to make things work. You are the client and you are there to witness the correct operation of the systems. Do not be afraid to play with the sensors and systems to see if they actually work: switch power on and off and see if it does exactly what is says on the tin.

BMS and aM&T witnessing

To witness BMS or aM&T systems, and if you have not dealt with these specialist systems in the past, invite a specialist. These systems are complex. Only a good specialist from your team will be able to make comments and make sure that the systems are commissioned and running as per design.

For BMS, use the commissioning engineer who has written the software for these particular systems (Schneider Electric, Trend etc.). He will witness the operation of the systems against the description of operation. Beware: the commissioning and witnessing of the BMS will inevitably involve systems shutdowns, so communicate this to estates and facilities.

Six ways subcontractors cut corners
without you noticing it

1.   Hiding services behind the ceiling

Ceiling fixers will become your worst nightmare. They work towards their own programme, and it's a programme that has nothing to do with your project. Most of them are on a price, so they just want to crack on with things. The ceiling installation contractors are not a part of your M&E team; they do not attend your progress meetings and can work for another contractor or for the client; they do not stop when the M&E program slips: they carry on installing ceiling grids and bars regardless.

The suspended ceiling, or sometimes even solid ceiling with no access, will certainly be in before M&E commissioning starts. Access to the FCUs, chilled beams, radiant panels, control valves, isolation valves, pipe work and cables will be severely compromised.

Delays with getting access equipment will become one of the reasons why the commissioning is always carried out after the practical completion and a long way into the time of beneficial use. By beneficial use I mean when services are operational and the building users use those services. However, the works have not yet been formally handed over to the client.

Expect to find valves with no insulation jackets hidden above the suspended ceiling. This will become a serious problem for CHW when water condenses in the summer months on cold surfaces, dripping through ceiling tiles resulting in complaints about flooding. Do not be surprised to see cables without containment being clipped to anything available. When the ceiling grid is up, it is impossible to do a good job with limited access. Expect untidy pipework and ductwork installation, and insufficient bracketing.

If the services are distributed in the ceiling void, the installation will be completed to a satisfactory level that will be just about functional. For the most part, this is not a fault of the installation team: how can you work through 600 x 600mm ceiling tiles, pulling cables or lifting ducts and pipes? First of all, this is not safe. Secondly, you cannot complete the job to the standard you want due to access and spatial limitations.

2.   Work completed with risk to H&S

There will be plenty of work completed in a way that is non-compliant with H&S law and best practices. This will involve serious H&S risks to the operatives. It will be frustrating, but contractors will be saving money by not doing the following:

⊗ Not providing adequate access equipment;
⊗ Not providing PPE adequate for the job;

⊗ Cutting essential training, such as: use of harnesses, asbestos awareness, PASMA, first aider, fire marshal and Site Safety Management Training Scheme (SSMTS);

⊗ Not keeping H&S, waste and environmental management, and fire and emergency paperwork up to the standard required by law;

⊗ People will be working with generic risk assessments and method statements. Operatives will be carrying out tasks outside of their competency zone because they are told to get on with it.

The sad thing about this is that people's health is put at risk. The small contractors on refurbishment projects are simply not used to carrying out works in a way that is required by the regulations. I know you think that you will make them comply, but it is like making your child go to bed early. I worked as a principle contractor and subcontractor, so I know the game inside out. Here are some ways the contractors cut corners on H&S:

1. Appropriate edge protection is expensive; therefore, contractors often use harnesses to work at height. The thing is that the operatives must be trained to use these harnesses. This means attending a course and keeping a register of harnesses and formally inspecting them weekly.

2. This also means that no lone person can work at height. How often do you see people working from step ladders and ladders on his or her own?

3. Edge protection often disappears before all the valve jackets are installed and the systems are pressure tested, commissioned, snagged and witnessed. Scaffold hire is expensive and is normally removed prematurely.

4. Instead of using a scaffold tower or a permanent scaffold, installers use stepladders and ladders to access the work. Having a mobile scaffold tower also means that those assembling them have to have PASMA tickets. The only way to make sure that scaffold towers are assembled, adjusted and used properly is to have all the mobile scaffold users trained. Again, training costs money. It also means people off work. It also means loss of production and waiting for the next available training opportunity plus travel expenses with accommodation.

5. Appropriate PPE is not issued. We are talking here about some basic items, such as appropriate gloves, bump caps, hard hats, ear defenders, dust masks, safety glasses and goggles.

6. The contractors do not have enough fire marshals or first aiders on site to cover colleagues who are on holiday. In fact, some contractors don't have any first aiders or fire marshals.

7. No Appointed Person (AP) for steam and electrical isolations. Contractors involved in electrical isolations and switching on power need to have a competent person for such works. This involves specific comprehensive training. The same applies to steam isolations.

8. No permits issued for working at height, confined space, electrical isolations, steam isolations and switching on power. The person issuing the permits needs to be available in the site office for those tasks. He also needs to lock the circuit breakers and steam valves. The AP also needs to re-inspect the completion of the works before he removes the locks and closes off the permit. All of these take lots of time and becomes a part-time job for someone. But contractors do not price for APs; therefore, these procedures are often not followed.

3. Damage each other's work, existing installations and customer's property

Nothing will stay in the way of the contractor completing their works. If they have to stay on existing pipes and damage the cladding, so be it. They will be too lazy and too rushed to engineer an alternative method of works. If they need to weld on the felt or asphalt roof, they will put one or two boards underneath the welding bench and will crack on with hot works. Sparks, swarf and even the remains of hot welding rods will be dropped on the felt, damaging the client's roof.

Rivet nails will be dropped by the laggers straight on the felted roof; when it gets hot in summer, swarf and rivet nails will cut through felt like a hot knife through butter.

Often, the contractors flood the areas they are working in and everything on the floor below, but never accept that they caused the damages. They say: "*It was like that before we started, it was like that all the time, it was another contractor, we were not in the area at the time*". They lie. At first it was hard for me to understand why. Now I understand.

It is a culture. You smoke on site and you get away with it, you do not wear the hard hat and you get away with it, you damage the client's equipment or property or someone else's work and you blindly lie.

You have to vet the contractors properly to get the responsible and responsive ones. You communicate the site rules in the contract and do weekly H&S workshops and toolbox talks in the working areas, talk to the installers, listen to them and explain why they need to do things in a certain way. It takes lots of your time but there is simply no other way.

4. Use existing containment to install cabling or not use it at all (bow and arrow job)

The BMS, electrical, fire alarm, communication and IT contractors will install the cabling using the existing and already overloaded trays. They will clip cables to pipework, ductwork, existing equipment, making the cable runs as short as possible. The IT installers and fire alarm cable installers are the worst offenders. It is often joked that they will tie the cable to the end of an arrow and shoot it using the bow. This will be the cable route from A to B. Then they cable-tie the cables to anything which is nearby and that's the end of the installation.

They often say that they will do containment later, which is an unbelievable excuse that would insult the intelligence of 3-year old schoolboy.

5. Leave redundant services without stripping them out

There are plenty of ducts, pipework, insulation, cables, trays, conduits, old switchboards, isolators, sensors and fans left in the lofts, ceiling voids, floor voids, roofs, plant rooms and other areas which are difficult to access. In some instances, it is simply impossible to get the stuff out in one piece; in others, it's plain laziness. Why do you need to sweat taking the stuff out if you can just leave it there?

Taking things out might be problematic in terms of safety as well. Once the services go through the wall, it might be hard or even impossible to trace where they go. This is why contractors often leave redundant services in situ.

6. Do not produce required handover paperwork, proper O&Ms and training

It is beyond comprehension how some small companies are not used to recording what they have done. This includes, updating wiring diagrams, charts, producing as-built drawings, data sheets for equipment, testing and commissioning sheets, snag lists, pressure test results, flushing, chlorination and balancing. This is one of the reasons why they are cheaper: they simply do not comply with H&S and environmental laws, or any building regulations, best practices or guidance notes. They simply do not allow for these in their prelims.

You will be using some client's preferred suppliers and therefore have to do three things: the first is to produce a draconian contract specifying all the items required of them. The second thing, and this again should become a part of the contract, is they have to allow for a design engineer and document controller to do those tasks. The last thing is that you have to allocate a separate budget for managing these things for your subcontractor. This is the only way it will work. Some small, preferred suppliers simply do not do things that are compulsory on any decent new-build project. They are not familiar with the processes and need help, education and patient coaching.

Site instructions, change of scope and variations

To get the project flowing, you will have to give plenty of site instructions (SIs) to your subcontractors. Some of them are contractually valid; some are not. However, having an instruction in the eyes of the subcontractor's manager is enough to get on with this particular part of his project. The more awkward, difficult, contractual and uncooperative the contractor is, the more site instruction requests you will issue.

If you neglect to issue site instructions, the subcontractor will threaten to stop progress. As the project manager, this is the last thing you need to happen. Consult the commercial support person for your project and issue the instruction. Often, the requests for the site instructions are contractually invalid.

When the time comes for adds and omits or to review the site instructions, your commercial manager or quantity surveyor (QS) will point out to the subcontractor's QS that this particular instruction does not have any financial impact on the project. Your QS will clarify that this SI is nothing more than an instruction to get on with works.

The principal contractor can employ a design consultancy to design the project and then get the subcontractors to build it. The other common option is doing a design and build (D&B) project where the subcontractors design their specialised systems, coordinate it and then build it.

The most important thing here is that a change of scope is much less likely to happen in a D&B contract than on a normal new-build project where the client or architect might introduce lots of changes. By the way, and just to clarify, the change of scope is a valid variation and should be accompanied by a site instruction. Normally, on refurbishment and retrofit projects, the scope is changed by one of these three:

1. As an afterthought, the design consultant makes a change to make something work or to cover certain regulations. For example, if there is insufficient flow through the newly installed steam plate heat exchanger (or if it is oversized), the consultant introduces a by-pass loop so that the plate does not lock itself out at high temperatures. If it is a design and build (D&B) contract, then there is no cost implication for the principal contractor; the D&B contractor will make all the changes without publicising them. If the contractor is not responsible for the design, then he will make a nice meal of advising you about the cost of rectifying something. As a principal contractor, you should always be aiming to pass on this cost to the designer.
2. Subcontractors introduce the changes to each other. The mechanical contractor makes changes affecting the BMS or electrical contractor, who picks these design changes instantly asking the principal contractor for the site instruction. The BMS contractor is also often found guilty of asking to cut an extra control valve or burn a pocket for the BMS sensor.
3. Changes by the client rarely happen in retrofit and are most certainly to do with maintenance. This happens when the client maintains the estate or if he outsources it to a facilities management (FM) company. In any case, the maintenance team might ask for n+1 redundancy or for certain elements of the specifications to be uprated. The other maintenance projects

running in parallel with your project might also affect the scope of your project. Always RFI those changes with the customer, as those RFIs will become a basis of your quests for the site instructions.

Depending on your relationship with your client, you can always try to ask for the site instructions from them. Depending on your approach and reasoning behind it, the customer might not want to cash out on the project, specifically, if there are no extras in the budget. However, it might be easier for your client to help you out by doing some tiny jobs for you via the maintenance budget. If the changes of scope are valid, the client will have to instruct you. You are unlikely to get extra cash on top of the project value, but you can offset some omits to the project introduced to you by the client or nature of works.

Recovery of works and acceleration

There are a few ways of clocking extra hours:

1. Work weekends;
2. Work a few extra hours into the evenings;
3. Working in shifts so there is another shift starting at, say, 17.00.

Sad practice shows that none of these work. People should rest from their job on weekends and recuperate. Besides, there are always engineering works on public transport during the weekend with the result that people get to work late and want to leave earlier because it takes them longer to get home. Weekend works need quality supervision. Your managers and supervisors are not robots and also need to recuperate after a stressful week, so they want to get home earlier as well. As soon as they go, the rest of the site team leave as well.

In addition, things are not quite organised during the weekend in terms of access, materials, design support, tools and coordination with the other contractors. In summary, you had a great weekend if your team delivered half of what they normally do during two normal working days.

The same applies to working, say, 11 hours a day Monday to Friday. People take breaks more often; they start hanging around with mates more often and try to find something to take their mind off their work. The job they like becomes a have-to-do-for-more-money thing. They stop having fun. They get home so late that all they want to do is eat, shower, and watch a bit of TV and go straight to bed. It takes its toll on the rest of the family.

Working in shifts is even more of a problem, and specifically if it involves more complex work . . . and by the way, all the works in retrofit are complex. Shift supervisors are unable to do an effective handover of working areas. The day shift had one idea of prefabricating the pipework, and the night shift takes over the works in a different manner with different tools, work practices and quality. It can get very frustrating at times. The most important thing is that shift work is unproductive and very, very expensive. If the night shift delivers a third of what the day shift normally delivers, it was a good night.

There is lots of hot talk about acceleration at management meetings. It is some sort of 'no brainier' and common-sense way of recovering the works so the program looks good. The thing is that this is where all management repeats the same mistake.

Once you slip on the program, the lost time is not recoverable. You can make stupendous investment in trying to accelerate progress, but the fact is that the most well-managed acceleration will only keep you from falling further behind. Acceleration can only work with very basic and repetitive works like pipework insulation.

To manage a slipping program, you need to look into the cause of the real problems. The best way known to me for identifying the problems is by following the works management system described earlier. You need to resolve the real problems and not waste your resources on a counterproductive acceleration of works.

The ineffectiveness of overtime does not apply if the works involve systems shutdowns. This is when out-of-hours time slots are allocated by the client for works completion. These works simply have to be done to continue in making progress. This actually works very nicely with your workforce, as they understand that there is no other time the work can be done. The benefits for the operatives are:

a)   The guys feel important, trusted and valued completing such critical tasks;
b)   They are assigned with a job and knock, i.e. they do it and then go home.

How practical completion is unpractical

Practical completion (PC) is nothing more than a line on the sand close to the project completion. Look at it this way and nothing more serious than that.

Until the systems are handed over, the client is enjoying the beneficial use of the systems. Beneficial use is the period when the system is operational and the building users enjoy the benefits from its operation. Normally, this applies to systems that are not fully commissioned. Practical completion is the next stage of the project when the systems are practically completed and functional.

However, there are still snags and some paperwork still outstanding. Handover is the final process, which often happens gradually, when the customer takes over the responsibility for the systems. This normally happens when the systems are fully installed, commissioned, de-snagged, all the O&Ms and as-built drawings are in place, and all the training and any other outstanding issues are resolved.

Big projects tend to drag on after the practical completion and even the handover. Contractors' management sit on the project review meetings and plan how to maximise the profit or reduce the loss on the project. Usually, this means bad news for the project team. Resources, like key personnel, are moved to other jobs. Temporary agency staff such as the site manager, document controller, administrator, H&S advisor and site labourer are off-hired too soon. Equipment, like cabins and H&S gear, are off-hired too. This is often a premature and short-sighted direction from the company's management.

As a result, the people resources left on the project are overloaded with work and are simply unable to simultaneously cope with dozens of issues and problems. There are no staff on the project to whom work can be delegated. Certain smaller issues are swept under the carpet or simply closed.

Towards the end of multimillion-pound projects, the drain on resources starting from operatives and management causes projects to slow down. Progress becomes unbearably slow. Less and less time is spent on quality control, commissioning, testing and snagging. Witnessing disappears from the project's vocabulary; less and less time is spent on paperwork, O&Ms and H&S. Design consultants disappear from the project, as they have already overspent on the account.

Little unfinished things tend to drag on. With limited labour on site, there are simply too many other things to complete. Lack of commissioning and witnessing causes systems to malfunction. New problems to do with the site's existing conditions and systems design surface. All of these issues and problems need resolving, but the contractors' management want you to handover the project and walk away from it already.

Design companies and commissioning management consultants disappear from the project to save costs. The customer calls every day or sends emails reporting issues with the newly refurbished systems and CC'ing twelve people. Complaints from the customers continue to come in. The maintenance team complains to the project team and you end up back on site solving these problems. Getting your contractors back is a hard task.

As a project manager, you have to be very strong, diplomatic and influential to be able to defend the resources on the job until you are confident enough that the project is over the line.

The best time to practically complete the job is when the following is done:

- ✓ Systems are fully commissioned and witnessed;
- ✓ Works are de-snagged by the client;
- ✓ O&Ms are reviewed and revised;
- ✓ Maintenance staff are comprehensively trained on the new installations.

Although, the equipment installed is under warranty for a year or so, the maintenance team is in charge of the operation of the systems. All the problems to do with the operation are their problems now. Sometimes, it is difficult to get across to the customer that you are no longer technical support for the project.

The customer does not have a systems support contract with your company. You have been helpful to the customer for the duration of the project; by default, the customer calls you because you have the answers. The maintenance team will be very skilful at pushing back any issues with the new systems to you. Some of the issues will be valid, but the majority will have nothing to do with you. They need to recruit a competent resource to look after the refurbished or new systems, not keep coming back to you as a contractor.

I would like to share with you a very effective way of finishing unfinished business with the contractors. By unfinished business, I mean completing 100% of the snags, acceptance of the counter charges you pass on from the client or other contractors, cleaning the site, handing over fully completed O&Ms, arranging the contractor to do proper training for the client and maintenance, obtaining .dwg type record drawings and having all the commissioning, witnessing and testing paperwork in place.

This method is effective because it works with even the most contract-obsessed, uncooperative, difficult, incompetent and awkward contractors in the world. During the project's duration, the contractor might have caused you an unbearable amount of grief and trouble, but now you will get you own back by withholding at least 15% of their contract value as a retention. They submit the application for payment to you and you don't pay, substantiating it by any unfinished business you have with them.

In front of your eyes, a magic transformation will happen, and an ugly duckling becomes a beautiful and very nice swan. All of a sudden you will get the treatment every client deserves. The contractor's project manager will be on the phone constantly trying to talk to you in order to meet up and agree the outstanding issues. Have a list of any outstanding items before the handover can occur. Issue the list before the meeting and talk it through your subcontractor recording the agreements, dates and actions.

In a week or two, review the list again, making sure that you are satisfied with what was done on site. Review the as-installed drawings and O&Ms. Make sure that training has been completed to a good standard. Check with your client that all the snags were de-snagged, areas 100% cleaned and that the other contractors do not have any claims against you which you can pass on to the other subcontractors. Make sure that your customer fully accepts the project from you first before accepting it from the subcontractors.

The most fulfilling thing for me on any project, apart from my systems functioning perfectly and customer satisfaction, is sitting with my client and signing the completion certificate. For me, that is a pinnacle of the project and a very special and bonding moment.

37

O&Ms - paperwork or a building guide?

Operation and Maintenance (O&Ms) manuals for retrofit projects are much more important than manuals for new-build projects. On new-build projects, the O&Ms are normally put together right at the end of the project only after the money for the last application for payment was withheld. That is normally when bells start to ring. It is considered an administrative task and often done by people not very competent in the project, because the majority of the project team have been transferred to other projects.

Normally, documents such as the data sheets with maintenance sections and as-built drawings are piled together. The systems operation manuals are written in a very basic manner just to get away with it. You see, the project team delivering new-build projects are almost 100% clueless about the maintenance of these systems.

To be a successful designer and contractor, one has to, firstly, understand how the buildings are maintained and what's involved. Secondly, design, build, commission, hand over the systems, train the maintenance team and the occupants, and do the O&Ms in a way to ease up the operation and maintenance of the building.

One way of improving maintenance is to produce quality O&Ms. Maintenance engineers, technicians and design consultants will have access to all of the information regarding the kit, maintenance and operation of the systems they need. For example, if the plate heat exchanger reaches a critical temperature and locks itself out, the maintenance engineer can open the installation and maintenance document for that unit, find the fault reference number along with a description for rectifying the code.

If the engineer is still in doubt, there is a manufacturer's technical support phone number cited; if the engineer calls, he will get answers to any additional questions. The O&Ms will also have a BMS section stating the operational set points and control strategies in place.

Agreed O&Ms will have:

1. Systems overview—this should be written in plain English;
2. Operation and maintenance data sheets for each piece of kit installed. Unfortunately, you often get sales literature or normal data sheets in the O&Ms;
3. As-built layout drawings, schematics and .dwg format files (soft copies) and technical submittals;
4. Manufacturer's and supplier's contact details;
5. Troubleshooting or reactive maintenance, e.g. what to do when things breakdown;
6. Routine maintenance;
7. Schedules of equipment like pumps, pressurisation units, plate heat exchangers, water softeners, AHUs, fans, etc. in a format of the asset list.

Great O&Ms will also become the basis for training, which you will learn about in the next chapter.

You can easily outsource the administrative task of putting the O&Ms together to someone else or to a specialist company. However, the most important thing is that you, your team and maintenance team are on a distribution list for comments and reviews.

Ask your client for a big meeting room and run one to two sessions on how to use O&Ms. People will appreciate it a lot, and you will learn even more about their training needs.

The H&S file is supposed to be completed by CDMc; however, it is beneficial if you recruit the same O&M specialist consultancy to complete it.

Training

You will need to train the maintenance staff and building users to operate and maintain the buildings the way you have set them up. Develop a training matrix for the organisation with your client. The matrix should include the names of people and what training they require. You cannot train everyone on everything, so the training matrix should be a balancing act between their needs and your budget. With the training itself, you will face the following challenges:

- ⊗ People do not want to be trained or even bother making the effort to learn;
- ⊗ Maintenance are not used to maintaining the buildings properly: why should they start doing it with your systems?
- ⊗ People cannot comprehend what you are training them on, or they are not allocated sufficient time to learn and play with the systems due to the operational constraints of the functional business;
- ⊗ People do not see the link between maintenance and energy savings. Their job spec is to have the kit running to minimise complaints, not to save energy.

Let's have a look at those constraints in more detail. The Facilities Management (FM) industry is guilty of low pay, low morale and high staff turnover. The mentality of many maintenance engineers and technicians is to clock in and clock out; their minds are elsewhere, and they are simply there to do a job. Your aim is to demonstrate that maintaining and operating newly refurbished or installed systems is fun. You have to direct these people so they understand that, from now on, the systems are theirs and they are now the masters.

If they understand how the system works, they can fix it. It is no longer outside of their competency zone.

Due to budgetary constraints, the majority of organisations, and specifically those in the public sector, reduce their maintenance budgets substantially. Maintenance becomes reactive, e.g. if something breaks, then they do something about it. Servicing the pumps, AHUs, plate heat exchangers, lighting ballasts, cleaning strainers, cleaning filters on FCUs and split systems, cleaning ducts, checking valves and accessories becomes non-existent.

In turn, this inevitably reduces the life of the equipment and parts; it also causes them to run inefficiently thereby wasting energy. Maintenance engineers are assigned jobs to sort out complaints from other departments; they simply do not have enough hands for preventative maintenance tasks. In the case that they do, it is mostly a quick look around to make sure the thing is still there and not leaking. This is just to tick the box and to close the job.

Another part of the problem is that maintenance personnel are not often competent for the tasks given to them. It is your job to become their personal trainer for a few sessions until they fully understand your systems, how they work, why they are set up in this way, what to do if components break and what happens if they start changing the settings. The consequences of them not operating the systems efficiently are sliding back to higher energy bills, excessive wear and tear and increasing costs of building maintenance.

Normally, the standard response to the 'too cold' complaint is to override the temperature setting, move the VT curve up, or get the equipment to run non-stop. All of these are hugely detrimental to saving energy. In fact, your project might even get a bad reputation if people become careless with settings. If a complaint comes in, log the temperatures on the BMS or go and measure it with a hand-held device. If it is 21 °C (for most of the occupancy types), do not change anything. Simply educate the person who raised the complaint that this is way within standards. Admin blocks are normally the worst offenders for temperature complaints.

If you see that any member of the maintenance staff does not fully understand you during training, slow down and know that you need more sessions. The best way of training is in the plantroom with a schematic drawing. Try to make it very practical for them.

> Go through the sequence of starting and shutting the plant in the plantroom, because this is one of the most common things the operatives will be carrying out;
> Cover routine and reactive maintenance;
> Fault finding;
> Isolation points;
> System components;
> Always emphasise the importance of using the BMS to understand what caused the problem in the first instance;
> Where to find information in O&Ms.

The maintenance team's awareness needs to be raised, often slowly, so that when they react to complaints, they make the right choices.

You may end up sounding like a head teacher at school, yet many people are clueless about the operation of the systems, as the systems are still shut and then started manually.

Arrange separate training for those people using the BMS. If required, limit or eliminate the options of changing something on the BMS for those personnel less competent.

You need to split electrical and mechanical training. This is because mechanical people are not normally assigned electrical fault-finding and vice versa. Arrange the same training in two slots. This will allow all the operatives to attend the training when their workload allows them to. If the systems are complex, then start with familiarisation in the plantrooms. Then, move on to operation of the systems, routine and reactive maintenance. Take your time and let people ask questions however silly these questions might seem to you.

Bear in mind that you spent months, if not years, on that job and know it like the back of your hand. For others, it is something completely new and it will take time to learn about it. Quiz the people at the end of training.

Up selling additional works, becoming a trusted partner and preferred supplier

Throughout the project, your customer might ask you to do some extra work. He will ask you to quote the job first to make sure that you are competitive. Because of your presence on site, and if the customer is satisfied with your project, you should get those little projects. Extra work might include changing some valves, lagging additional pipes, installing some extra lighting, small-scale power distribution, BMS upgrade or installing some TRVs. Contract-wise, these extra works can be treated as adds and omits.

Large organisations spend tens of millions of pounds on annual maintenance. Although it can be tough for them planning for additional projects, the reality is that reactive projects happen non-stop. For many preferred small-scale suppliers, it is like a feeding table. Little maintenance projects of up to £10k are produced non-stop. These preferred small-scale suppliers are cheap and they deliver a reasonably good quality of work. They are on the site all of the time sorting out bits and pieces. They know the site extremely well; they also know the customer's organisation inside out.

You should aim to become the preferred local supplier for these works. Throughout the project, the customer learned much about your team. If you delivered consistently well during the project, and have taken care of the customer's pains, there is no reason why you should not be getting additional work. You see, if someone performs well, we become accustomed to the good performance; if we lose this contractor at the end of the job, it is actually a very painful experience—we get used to the easy life.

All you need do is be very competitive on the price, even though it might be not your final decision about the gross margin of the project; however, it is certainly an area over which you have influence. At the end of the day, gross margin is a balance between profit and the risk of making a loss. If you can convince your management that there is no risk due to your site knowledge, and being so close to the customer's heart, then you will open the gate for a constant flow of small projects for your organisation.

If you deliver the project well, and the customer is happy, then it is likely that you were happy on that job as well. You should aim to have customers for life by staying and delivering more projects on site.

Ask the director of estate how they want to budget for the additional (phase-two) project. They might want you to wrap phase-two into another multimillion-pound project and produce a winning business case for the board. The director of estate might also have separate pockets of money he wants or can spend on energy reduction projects.

By the way, the majority of M&E work can be focused on energy reduction. Replacing unreliable boilers or steam calorifiers reduces energy bills, reactive maintenance and backlog maintenance for the organisation; it also substantially reduces the risk of equipment failure and business disruption.

Get assistance from your best sales guy or business development manager to help you wrap it all up into a win-for-all-parties project.

Another great option is selling capital projects on a basis of the medium payback (4-7 years) via the energy reduction the project achieves. If you want guaranteed energy performance, then implement an energy performance contract. You can read about the energy performance contract in my book, *Energy Performance Contracts Handbook*. Also, find out about how to come up with the energy conservation works by studying my third book called *Energy Audit UNDONE*. Both are on Amazon in hard and soft copy.

You can also help to raise funds through energy reduction incentives available from the Government to leverage your sale.

Afterword

There are a great number of engineers with no practical experience, knowledge or application of their engineering labour. There are managers who have no elementary understanding of what they are managing and no skills in management; they do, however, have plenty of confidence and bravado. And then there are designers who have never been on site, have no understanding of maintenance and no appreciation of the specialities outside of their immediate niche.

The funny thing about the construction industry is that many design consultants are clueless about maintenance, and they never find out if the project was installed on budget; the majority of them never witness commissioning. This begs a question: how can you make sure that the building is doing what it is designed to do if you do not witness the system's operation in action?

And then there are great mechanical and electrical (M&E) project and site managers who can complete a new-build M&E installation, but they are clueless about the design, commissioning, training and maintenance of the M&E installation.

As the result or the skills misalignment, building services systems:

    i.   Are never designed for energy efficiency, commissionability and maintainability;
    ii. Run in some sort of fashion, not as per design;
    iii. Run inefficiently;
    iv. Are unmaintainable;
    v. Expensive to run;
    vi. Take too much time to install and commission;
    vii. Not practical to operate.

It is the job of the outstanding project manager to make sure that this doesn't happen.

Masterful project management is about delivering the project, making a customer happy, making money on the project (this where you get your bonus), happy building users and creating great systems for maintenance engineers to work with. It is also about rejuvenating assets and systems that were falling to pieces and granting them a new lease of life. Masterful project management is also about training the maintenance staff and building users on how to operate and maintain the systems that you put in place.

Outstanding project management is not excessively loaded with unproductive installation, commissioning and progress meetings. It is not aggressively driven by unachievable deadlines. Successful projects are not micro-managed by senior management when things slip up. The main theme here is to install great quality systems and commission them in a way that maintenance people can enjoy running them.

At the same time, all retrofit and refurbishment should be energy focused. This means having thought-through design and making sure that the building services are commissioned and maintained with energy efficiency in mind.

Some people contribute to the charity of their choice or volunteer to help the community. I write books. I invest a substantial amount of my personal time writing books, as I aim to raise awareness of building projects, sustainability and the construction industry overall and raise the industry's ethics.

By publishing this book, I've coached you and helped you to learn and polish new, desirable skills that are essential for the project's success. In return, I have two final requests for you:

1. Please leave a two-to-three sentences review on this book by following the link below to Amazon: http://goo.gl/Okfo9j
2. Now that you've grasped the benefits of my site management book, I would like to ask you to buy at least one of my other books right away. Both books are available from Amazon:
   ✓ Energy Audit UNDONE
   ✓ Energy Performance Contracts Handbook

It has been my pleasure sharing my learnings with you, and I ought to thank my editor, Michael Griffiths, for making sure that the knowledge this book provides is easily digestible and presented in a structured and intuitive way.

. . . till our paths cross again, my Jedi . . .

Sincerely yours,

Viktar Zaitsau

Appendix A

Examples of building services retrofit works
Energy generation (energy centres)
Retrofitting CHP and boilers
Boiler replacement
Water treatment and filtration
Renewable and low carbon technologies
CHP
Tri-generation
PVs
Solar thermal
Ground source heat pumps
Air heat pumps
Wind generation
Biomass
Building fabric
Installation or improvement of loft, cavity wall or solid wall insulation
Reflective panels behind radiators
Draught proofing (windows, doors and roller shutters)
Double and secondary glazing
Solar films
Domestic Hot Water (DHW)
Straight replacement of the DHW system, fitting water softeners
Fitting points to use electric heaters for buildings with very low DHW use
Eliminate the DHW storage by fitting plate heat exchangers
Flow reducers/PIR sensors/push taps on DHW and general plumbing works
Heating and pipework
District LTHW (heating) system
Converting steam distribution to LTHW
Re-balancing the radiators and heating system

| TRVs for radiators |
| Water treatment side filtration units |
| Refurbishment and replacement of the heating distribution system |

| **Swimming pools** |
| Installing new plantroom and new ventilation and air conditioning systems |
| Installing biomass, CHP and solar thermal as it is steady constant load |

| **Compressed air systems** |
| Compressor replacement |
| Controller for sequencing and pressure reduction at night and weekends |
| Repair leaks |

| **Cooling** |
| Free cooling using blast chiller |
| Chiller replacement, data centres and cooling upgrades |
| Evaporative cooling units for data rooms (needs water and drain) |
| Free cooling of the IT rooms using supply and extract fans instead of DXs and air source heat pumps |

| **Pumps and fans** |
| Replace old pumps with direct on shaft energy efficient once |
| Install EC motors with backward curved fans |
| Install VSDs with pressure control of the static pressure sensors |

| **AHUs and ventilation** |
| AHUs replacement or refurbishment |
| Ductwork extensions and alterations |
| Heat reclamation: thermal wheel, recuperator, re-circulation and run around coils system |

| **Electrical** |
| Install or repair power factor correction equipment an harmonic filtering |
| Install low and super low-loss transformers |
| Systems refurbishment and replacement |
| Voltage optimisation/reduction |
| PC shutdown/servers optimisation software |
| Electrical distribution systems upgrade |

Lighting
Occupancy and Lux level control or installing brand new system
Replacement with high efficiency fittings
Sun Tubes and skylights

Water
New DCW (Domestic Cold Water) system
Flow restrictors, push taps or aerators
Waterless urinals or cistern filling controls
Rain harvesting or recycling as 'greywater'

General
Upgrade lifts
Install metering system and aM&T (automated monitoring and Targeting)
Repair leaks: steam, condensate, LTHW, CHW, DHW and DCW
Insulate steam, condensate, LTHW, CHW and DHW pipework. Valves, equipment, flanges and accessories to be insulated with jackets

BMS upgrade and optimisation
Upgrade standalone control
Sequence control of the boilers
Eliminate requirement for humidification and de-humidification
Replace passing LTHW and CHW control valves
Install additional zone control valves
Occupancy sensors for AHUs
VSDs on to AHU fan motors
Chiller sequencing control
BMS optimisation: optimum start/stop, night purge. Increasing the dead band for summer /winter, etc.
Upgrade of BMS controllers